AF764719

NUTRITION FOR FOODSERVICE AND CULINARY PROFESSIONALS

Nutrition for Foodservice and Culinary Professionals

Editor

Oliva Cabral

scitus academics

Nutrition for Foodservice and Culinary Professionals
Edited by **Oliva Cabral**

ISBN: 978-1-68117-552-2
Library of Congress Control Number: 2016942959

© 2017 by
SCITUS Academics LLC,
www.scitusacademics.com
Box No. 4766, 616 Corporate Way,
Suite 2, Valley Cottage,
NY 10989

This book contains information obtained from highly regarded resources. Copyright for individual articles remains with the authors as indicated. All chapters are distributed under the terms of the Creative Commons Attribution License, which permits unrestricted use, distribution, and reproduction in any medium, provided the original author and source are credited.

Notice

Reasonable efforts have been made to publish reliable data and views articulated in the chapters are those of the individual contributors, and not necessarily those of the editors or publishers. Editors or publishers are not responsible for the accuracy of the information in the published chapters or consequences of their use. The publisher believes no responsibility for any damage or grievance to the persons or property arising out of the use of any materials, instructions, methods or thoughts in the book. The editors and the publisher have attempted to trace the copyright holders of all material reproduced in this publication and apologize to copyright holders if permission has not been obtained. If any copyright holder has not been acknowledged, please write to us so we may rectify.

Printed in United States of America on Acid Free Paper

PREFACE

Nutrition is the science that interprets the interaction of nutrients and other substances in food in relation to maintenance, growth, reproduction, health and disease of an organism. It includes food intake, absorption, assimilation, biosynthesis, catabolism and excretion. Beyond the physical and emotional satisfaction of eating, we often are concerned about how food choices affect our health. Eating a healthy diet and exercising is not just a concern for adults but for children and teenagers as well. An attribute of a food that includes its appearance, smell, taste, feel in the mouth, texture, temperature, and even the sounds made when it is chewed. The nutritional content of a food can be an important factor in deciding what to eat. Knowledge of nutrition and a positive attitude toward nutrition may translate into nutritious eating practices. Just knowing that eating lots of fruits and vegetables may prevent heart disease does not mean that someone will automatically start eating more of those foods. A poor diet may have an injurious impact on health, causing deficiency diseases such as blindness, anemia, scurvy, preterm birth, stillbirth and cretinism; health-threatening conditions like obesity and metabolic syndrome; and such common chronic systemic diseases as cardiovascular disease, diabetes and osteoporosis.

Nutrition for Foodservice and Culinary Professionals is intended for students in culinary programs, as well as those in hotel, restaurant, and onsite management programs. Practicing culinary and management professionals will find it useful as well. This book helps readers use nutritional principles to evaluate and modify menus and recipes and to respond to customers' critical questions and dietary needs. From students in culinary arts, hospitality management, and nutrition and dietetics programs to practicing culinary and management professionals, this book will be an invaluable reference.

TABLE OF CONTENTS

1

PREPARATION AND EVALUATION OF GRANOLA – A BREAKFAST CEREAL, SUSTITUTED WITH MAIZE (ZEA MAY) AND COCONUT (COCOS NUCIFERA) BLEND

Eke - Ejiofor Joy*, Beleya Ellen Aswei, Gbarasogo Mbarabari Nicholas

Department of Food Science and Technology, Rivers State University of Science and Technology Nkpolu, Port Harcourt, Nigeria

ABSTRACT

Granola was produced using three varieties of maize namely, yellow, white and pop and oat as control. Coconut was added to the granola sample in order to replace walnut in the control. They were subjected to sensory, chemical and functional evaluation. Sensory analysis showed that there was significant difference in their color and taste, while flavor, texture, crispness and overall acceptability showed no significant difference ($p \geq 0.05$). Chemical analysis result showed that oat based granola had the highest value for fat, protein, energy, sugar, starch and amylopectin, while yellow maize granola had the highest value for moisture content and carbohydrate and white maize granola had the highest value for crude fibre and amylose. Granola produced from pop maize had the highest value for ash. There was no significant difference ($P \geq 0.05$) in ash, protein energy and amylopectin between oat granola (control) and maize based granola samples. Functional analysis of the samples showed that oat granola had the highest value for water absorption capacity and swelling power. Yellow maize granola had the highest value for dispersibility and bulk density. Pop corn granola had the highest value for solubility. There was no significant difference ($P \geq 0.05$) in bulk density and water absorption capacity.

KEYWORDS

Granola, Substitution, Preparation, Evaluation, Maize, Coconut, Breakfast - Cereal

1. INTRODUCTION

Granola is a ready-to-eat breakfast cereal that is made from grain typically high in carbohydrate, low fat and often a good source of fibre (Annelisse, 2009). It is a breakfast food made from oats, walnut, peanut and wheat which is usually baked until it is dry and crisp. A breakfast cereal therefore can be said to be a ready to cook, ready to eat convenient food product. They include puffed, flaked, shredded and granular products made from wheat, maize, oats, rice and barley. They may be enriched with sugar, honey, a malt extract and fortified with vitamins, minerals and nutrients (Annelisse, 2009).

Besides serving granola as food for breakfast, it is also eaten by those who are hiking or camping because of it lightweight and high calories. Granola is often eaten with yoghurt, honey, fruit and milk. It can also serves as topping for various types of pastries and/or dessert.

Oat and walnut are the main raw materials for the production of granola, which is relatively high in cost due to the fact that it is not grown in Nigeria. Granola is unpopular and therefore its consumption is very low and so stands the status of a smuggled product from Ghana where it is more of a staple food. There is little information in literature on the product granola and therefore this research.

Maize which is locally available and of low cost may serve as a substitute for oat and coconut for African walnut in the production of substitute granola. It is the most widely grown crop (FAO, 2009) and the fourth most consumed cereal in the past two decades, below sorghum, millet and rice (FAOSTAT 2012). Being among the primary food staples, maize consumption is widespread across the country and among households of different status. It is widely used in the preparation of traditional foods, and has two major types, white and yellow with a high content of carbohydrate, moderate protein (zein), fat, but rich in fibre, mineral and vitamin. Its protein contains low amount of gluten but a good source of starch which is used for baking and its oil used for cooking (Watson 1987).

In Nigeria, coconut is grown mainly for food and wholly eaten raw, until recently when it is being processed on a small scale into candies and chips (Asiedu, 1989). Coconut has various uses and its flour has been developed for use in baking to combat malnutrition (Grinwood, 1975). It is used as a source of sugar, dietary fibre, protein, antioxidant and vitamin (Paniappan, 2002). The coconut residue is made into flour (Timida *et al.*, 2001) and believed to contain dietary fibre, which has an important health implication in the prevention of risk of chronic disease such as cancer, cardiovascular disease and diabetes. Studies revealed that consumption of high coconut flour increase fecal bulk (Arancon, 1999). Coconut flour is gluten-free, though contain a significant protein. Its flour is used as composite in the baking of some product to improve the nutritional composition. According to Barret *et al.*, (2004) when coconut flour is incorporated into wheat flour, it increases the amino acid content especially lysine.

Peanut flour/paste remains underutilized and research is needed to develop new value added products from this raw material that is mainly used for livestock. Peanut has assumed significance in the recent years as a protein source in diets due to its high proteins content (25.80%). Its meal can be dried and ground into a paste form that can be added to various daily consumed foods (Zhao *et al.*, 2011).

Continuous search for wholesome food and poor nutritional quality of most readily available breakfast meal have lead to the utilization of other cheap underutilized and available food raw material which are desirable in reducing the use of high cost raw material such as oat and wheat to further enhance product development. This research intends to develop an adequate and nutritionally enriched breakfast meal to meet the need of the teeming population of our society. The objectives of this study therefore are, to prepare granola from locally available raw materials such as maize, coconut and peanut and to evaluate the sensory, chemical and functional properties of the granola samples.

2. MATERIALS AND METHODS

2.1 Materials

Maize (Zea may), oat (Avena sativa), peanut (Arachis hypogaea), coconut (Cocos nucifera), wheat (Triticumspp), milk, sugar, vegetable oil and vanilla flavor were Purchased from Mile 3 Market, while African walnut was purchase from "Spar" supermarket, all in Port Harcourt, Rivers State, Nigeria.

2.2 Chemicals

Chemicals used for this analysis were of analytical grade and were all obtained from the Biochemistry laboratory, Department of Food Science and Technology, Rivers State University of Science and Technology, Port-Harcourt, Rivers State, Nigeria.

3. METHODS

3.1 Preparation of Maize Meal

Maize grain (white, yellow and pop) varieties were sorted to remove spoilt grains. They were cleaned, winnowed and milled using a dry milling machine. The meal obtained were then stored in an air-tight container for use in the preparation of granola.

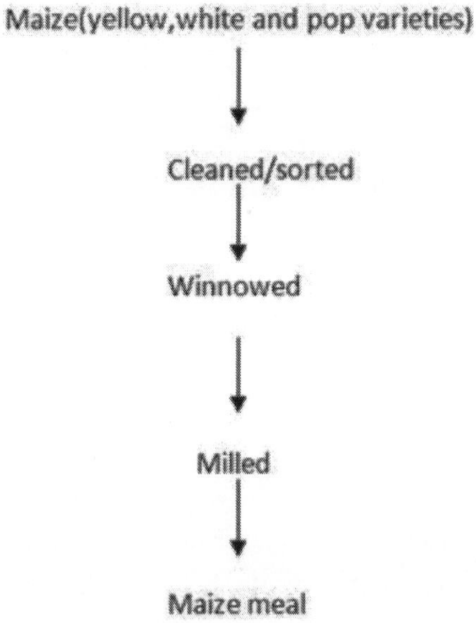

Source: Adapted from Houssou and Ayemor, (2002).

Figure 1. Flowchart for the production of maize meal.

Table 1. Recipe for the production of granola.

Ingredients	Quantities	
	Standardgranola (control)	Alternativegranola
Oat	500g	_
Maizemeal	_	500g
AfricanWalnut	80g	
Coconutmesh		80g
Groundpeanut	160g	160g
Wheatflour	100g	100g
Sugar	160g	160g
Water	200ml	200ml
Vegetableoil	16ml	16ml
Vanillaflavor	4ml	4ml

Source: George and Esther, (2003).

3.2 Preparation of Granola

Granola was prepared according to the method described by George and Esther, (2003). Maize meal, coconut mesh, ground peanut, wheat flour and sugar were weighed into a bowl and mixed. Water, vegetable oil and vanilla flavor were added to the weighed sample in the bowl and were mixed into dough kneaded and cut into shape. The dough was spread on the tray and baked in an oven at 130°C for one hour. The baked product (granola) was allowed to cool and subsequently store in an airtight container.

3.3 Sensory Evaluation

Granola samples were subjected to sensory evaluation within 24 hours of production. The granola was evaluated in two forms; dry and in milk solution the form in which it is to be served. A uniform quantity of granola and milk/sugar solution were given to all panelists in order to provide a base for assessment.

The dry sample was evaluated for color/appearance, crispness, aroma, taste, texture and overall acceptability while granola in milk solution was evaluated for color/appearance, taste, texture/mouthful flavor and overall acceptability.

A questionnaire was used to assess the above mentioned attributes using a 5-point hedonic scale with 5 = like extremely, 4 = like very much, 3 = neither like nor dislike, 2 = dislike very much, 1 = dislike extremely (Larmond 1977). Twenty (20) semi trained panelists drawn from within and outside the Department of Food Science and Technology, who are regular consumers of maize and coconut and who were neither sick nor allergic to any component of the raw material used for the production of the products at the time of the evaluation, were involved in the assessment. The panelists were instructed to rinse their mouth with water after tasting each granola sample.

3.4 Statistical Analysis

Results were statistically analyzed by using analysis of variance technique. Level of significance within means was calculated by using the Duncan Multiple Range Test (Steel and Torrie, 1980).

3.5 Chemical Analysis of Granola Samples

The chemical analysis of granola samples was determined using the AOAC (1990) methods for moisture, ash, protein, fat and fibre. Total carbohydrate was calculated by difference of moisture, ash, protein, fat and fibre. The total energy value of the samples was determined according to the method described by Mahgoub (1999). Amylose content was determined according to the method described by Williams *et al.*, (1970), while amylopectin was calculated by difference (% Amylopectin = % starch - % amylose). Starch and sugar was determined by the method described by Eke 2006.

3.6 Functional Properties

Dispersibility was determined by the method of Kulkarni *et al.*, (1991), while the relative bulk density of samples was determined by the method described by Narayana and Narasinya, (1984). The water absorption capacity was determined using the method by Sosulski, (1962) while swelling Power and Solubility was determined using the method of Takashi and Sieb, (1988).

4. RESULTS AND DISCUSSIONS

4.1 Sensory Evaluation Result of Granola

Table 2 shows the sensory evaluation result of dry granola samples prepared from three different maize varieties namely white, yellow and pop corn.

Color ranged from 2.00 – 4.15 with samples C and D (pop and yellow maize granola, respectively) as the most preferred, while taste ranged from 2.90 – 4.15 with sample A (control) having the highest score. Color and taste are important sensory attributes which to a large extent determines the acceptability of food products. In the present study color and taste showed a significant difference ($p<0.05$) between the control and the maize based products. Flavor ranged from 3.10 – 3.30 with sampleD (yellow maize) as the highest, while texture had 3.30 for sample A (control), B (white maize) and C (pop corn maize) and 3.55 for sample D (yellow maize), with no significant difference occurring in texture. Crispness ranged from 3.30 – 4.00 with sample C (popcorn) having the highest. Overall acceptability ranged from 3.35 – 3.60 with samples A (control) and C (popcorn) as the highest. Flavor, texture, crispness and overall acceptability showed no significant differences ($p>0.05$) amongst the samples. This suggests that acceptable granola can be produce from the combination of maize, coconut and peanut.

4.2 Sensory Evaluation of Granola in Milk/Sugar Solution

Table 3 shows the sensory evaluation result of granola samples consumed in a given quantity of milk and sugar in the ratio 3: 1, weight for weight. Color/Appearance ranged from 2.70 – 4.05 with sample D (yellow maize) as the highest, while taste, flavor, texture and overall acceptability ranged from 2.95 – 4.05; 3.15-4.05; 3.10-3.75 and from 3.15 – 3.85 with sample A (control) having the highest respectively.

Table 2. Sensory evaluation result of dry granola samples.

Sample	Color	Taste	Flavor	Texture	Crispness	Overall Acceptability
A	$2.00^a \pm 0.92$	$4.15^a \pm 1.14$	$3.20^a \pm 1.51$	$3.30^a \pm 1.45$	$3.45^a \pm 1.19$	$3.60^a \pm 1.14$
B	$3.30^b \pm 0.94$	$3.05^b \pm 1.09$	$3.10^a \pm 1.09$	$3.30^a \pm 1.22$	$3.30^a \pm 1.03$	$3.55^a \pm 0.99$
C	$4.15^a \pm 0.88$	$2.90^b \pm 1.16$	$3.25^a \pm 1.16$	$3.30^a \pm 0.98$	$4.00^a \pm 0.85$	$3.60^a \pm 0.95$
D	$4.15^a \pm 1.99$	$3.55^a \pm 1.31$	$3.30^a \pm 1.08$	$3.55^a \pm 1.28$	$3.30^a \pm 1.17$	$3.35^a \pm 1.35$

Means with the same superscript in the same column are not significantly different ($P > 0.05$).

Key: A= Oat granola (control).

B= white maize granola.

C= popcorn granola.

D= yellow maize granola.

Table 3. Sensory evaluation result of granola in milk solution.

Sample	Color	Taste	Flavor	Texture	Overall acceptability
A	2.70a±1.17	4.05a±1.09	4.05a±0.94	3.75a±1.02	3.85a±1.09
B	3.20a±0.95	3.40b±0.94	3.15b±1.18	3.10a±1.07	3.20a±1.11
C	3.50b±1.00	2.95b±1.09	3.25b±0.91	3.40a±1.27	3.50a±1.24
D	4.05a±1.24	3.05b±1.31	3.15b±1.34	3.10a±1.37	3.15a±1.18

Means with the same superscript in the same column are not significantly different ($P > 0.05$).

Key: A= Oat granola (control).

B= white maize granola.

C= popcorn granola.

D= yellow maize granola.

Table 4. Chemical composition of granola (%).

Sample	M.C	Ash	Fat	Crude Protein	Crude Fiber	CHO	Energy (Kcal/100g)	Sugar	Starch	Amy	Amylop
A	5.65b±0.21	1.47a±0.09	22.53a±0.68	10.92a±0.96	2.58b±0.14	56.58b±0.01	473.75a±6.02	11.17b±0.04	80.37a±0.15	40.11c±0.07	40.26a±0.07
B	6.60a±0.00	1.45a±0.02	20.69b±0.83	6.65a±0.72	3.91a±0.02	60.75a±0.08	455.32a±7.79	10.59b±0.0	74.29c±0.25	40.66a±0.07	33.63b±0.07
C	6.30a±1.28	1.53a±0.07	18.29c±0.06	9.74a±0.24	2.06c±0.06	60.08a±0.08	451.89a±0.40	10.84b±0.04	77.35b±0.50	40.27c±0.07	37.08a±0.07
C	6.75a±0.07	1.43a±0.07	19.38c±1.52	7.18a±2.41	3.70a±0.21	61.56a±0.00	450.19a±15.56	11.04a±0.04	76.25b±0.29	40.33b±0.15	35.92a±0.16

Means with the same superscript in the same column are not significantly different ($P > 0.05$).

Key: A= (control)

B= White maize granola

C= Popcorn granola

D= Yellow maize granola

M.C= Moisture content

CHO = Carbohydrate

Amy = Amylose

Amylop = Amylopectin

4.3. Chemical Composition of Granola

Table 4 shows the chemical evaluation result of granola prepared from three different maize varieties namely white, yellow and pop corn varieties. Moisture content ranged from 5.65% - 6.75% with sample A (control) as the least and sample D (yellow maize) as the highest. Moisture content of granola ranging from 5.65% to 6.75% is less than the finding of Annelisse (2009) with the 15.1% with use of resistant starch for production of granola bar. The low moisture content of these products indicates that the product would have an extended shelf life.

Ash content ranged from 1.43% - 1.53% with sample D (yellow maize) as the least and sample C (popcorn) as the highest. This result agrees with the finding of Jeffrey *et al.*, (2011) who reported 1.50% ash content for corn meal, while fat content ranged from 18.29% - 22.53% with sample C (popcorn) as the least and sample A (control) the as highest. Fat content of granola in this study agrees with the findings of Annelisse (2009) with a value of 18.60%. The higher fat content of sample A (control) may be as a result of high fat content of African walnut, which was substituted with coconut, thereby giving a product of less fat content. Fat content may also be influenced by varietal difference

Protein content ranged from 6.65% – 10.92% with sample B (white maize) as the least and sample A (control) as the highest. Protein content in this work agrees with the finding of Muhammad *et al* (2012) with a value of 9.00% of corn grit. The protein result of sample A (control) with a higher protein content is expected because of the presence of wheat which is known for its high protein, in relation to maize.

Crude fibre result of 2.06% to 3.91% agrees with the findings of Muhammad *et al.*, (2012) with a value of 3.91% for granola bar substituted with native syrup and dietary fibre. This result indicates that maize is a rich source of fibre.

Carbohydrate content ranged from 56.85 – 61.56% with samples A (control) as the least and sample D (yellow maize) as the highest. This result agrees with the findings of Annelisse (2009) with a value of 58.9%, while energy ranged from 451.89kcal/100g– 473.75kcal/100g with sample C (popcorn) as the least and sample A (control) as the highest. This result also is in agreement with the findings of Annelisse (2012) with a value of 450kcal/100g. The result shows that despite the change in the composition of the original ingredients, it did not adversely affect the derivable energy in the new product.

Sugar and starch contents ranged from 10.59% to 11.17% and from 74.29% to 80.37% with sample B (white maize) as the least and sample A (control) as the highest respectively. There were no significant differences ($P \geq 0.05$) in sugar and starch contents of the samples.

Amylose content ranged from 40.11% to 40.27% with sample A (oat granola) has the lowest value and sample B (white maize) has the highest value, while the reverse was the case for amylopectin which ranged from 33.63% to 40.26% with sample B (white maize) having the least and sample A (control) having the highest. Amlyose is the linear components of starch and it imparts definite characteristics to starch and therefore, its content is an important criterion in starch quality (Kurup, 1994). Amylose and amylopectin showed significant difference ($P \leq 0.05$) between the samples.

4.4. Functional Properties of Granola

Table 5 shows the functional properties of granola prepared from three different maize varieties namely white, yellow and pop corn varieties.

Dispersibility and bulk density ranged from 66.00%- 74.50% and from 0.40 – 0.47g/ml with sample A (oat granola, control) as the least and sample D (yellow maize) as the highest. Dispersibility in water shows the ease of break-up of agglomerates which allow particles to sink below the surface and disperse rapidly in liquid (Tizazu and Emire 2010), while bulk density is very important in packaging and material handling since it enables a higher amount of material occupy a smaller volume (Karma *et al.*, 1981). There was a significant difference ($P \leq 0.05$) between oat based granola and maize based granola in dispersibility and no significant different ($P > 0.05$) in bulk density amongst the samples.

Table 5. Functional properties of granola samples.

Sample	Dispersibility (%)	Bulk Density (g/ml)	WAC (g/g)	Solubility	Swelling Power (g/g)
A	66.00b±2.82	0.40a±0.01	1.04a±0.00	23.07b±0.49	5.89a±0.18
B	71.50a±2.12	0.46a±0.02	0.96a±0.01	23.89b±0.16	4.85b±1.16
C	73.00a±0.00	0.46a±0.00	0.95a±0.05	25.18a±0.16	4.57b±0.14
D	74.50a±0.70	0.47a±0.00	0.81b±0.00	25.15a±1.41	4.92b±0.01

Means with the same superscript in the same column are not significantly different ($P > 0.05$).
Key: A= Oat granola (control).
B= white maize granola.
C= popcorn granola.
D= yellow maize granola.
WAC = water absorption capacity.

Water absorption capacity ranged from 0.81-1.04g/g with sample D (yellow maize) as the least and sample A (oat granola, control) as the highest. Niba *et al.*, (2001) described water absorption capacity as an important processing parameter that has implications for viscosity. Furthermore, water absorption capacity is important in bulking and consistency of products. Increase in water absorption capacity in food systems enables end users to manipulate the functional properties of the dough in the bakery products. The yellow maize granola (sample D) was significantly different ($P \leq 0.05$) from the other samples.

Solubility ranged from 23.07%-25.18% with sample A (control) as the least and sample C (popcorn granola) as the highest, while the reverse was the case for swelling power which ranged from 4.57-5.89g/g with sample C (popcorn) having the least value and sample A (control) having the highest value. Solubility reflects the extent of intermolecular cross bonding within the granule (Hari *et al.*, 1989). In agreement with the above statement, Safo-Kantanka *et al.*, (1996) stated that the swelling power of a starch based food is an indication of the strength of the hydrogen bonding between the granules. Swelling power therefore is a measure of swollen starch granule and the food eating quality is connected with retention of swollen starch granules (Richard *et al.*, 1991). Swelling capacity is also a function of the product to rise when having interaction with water. Finney (1994) reported that the swelling capacity affects the temperature at which a product forms gel in maize flour; a similar observation was made in the present study. Solubility and swelling power showed significant difference between the samples.

5. CONCLUSION

Granola can be prepared from maize meal and coconut. The study has shown that maize and coconut can be substituted for oat and walnut in the production of granola of acceptable quality without altering the flavor, texture and crispness of the original oat based product. Maize based granola showed an increased value in carbohydrate, fibre and energy with energy showing no significant difference, while functional properties of the products also showed better dispersibility and solubility of maize based granola. However there was no significant difference in bulk density and water absorption capacity between the maize based granola and the control. Therefore maize and coconut can be used to produce an organoleptically, nutritionally and acceptable breakfast cereal granola.

REFERENCES

1. Association of Official Analytical Chemists (AOAC), 1990. Official Method of Analysis,17th Edition. Washington DC, USA.

2. Anneliss Eaigster (2009). Physicochemical and sensory properties of resistant starch based cereal products and effects on postprandial glycemic and oxidation stress responses in Hispanic women. *J Nutr.* 27:23-30.

3. Aranco, R. N. (1999). Asia-Pacific Forestry Sector Outlook Study: Asiedu, J.J., 1989. Processing tropical Crops – A Technological Approach. The Macmillan Press Ltd.

4. Barrett, A., Cardello, A., Maguire, P., Richardson, M., Kaletunc, G. and Lesher, L. (2002).Effects of sucrose ester, dough conditioner and storage temperature on long-term textural stability of shelf-stable bread. *Cereal Chemistry* 79: 806–811.

5. Eke J. (2006): Chemical, functional and pasting properties of starch and tapioca produced from selected hybrid cassava (manihot esculenta crantz) cultivars. Unpublished Ph.D Thesis.

6. Food and Agricultural Organization (FAO), 2009. Economic and Social Department: The Statistical Division.

7. Finney, K.F (1994). Contribution of individual chemical constituents to the functional (bread making) properties of wheat in cereals 78 better Nutrition for the World Million Minnesota. American Association of Cereal Chemist Inc.

8. George Pamplona-Roger and Ester Malaxetxebarria (2003). Recipes for Healing and Prevention. 3rdEdition in English.

9. Grimwood, B.E. (1975) Coconut Palm Products. FAO, Rome.

10. Hari, P.K, Gargs, S;Garys, S.K (1989).Gelatinization of starch and modified starch. *Starch* 41(3):88-91.

11. Houssou, P. and G.S Ayemor (2002). Appropriate processing and food functional property of maize flour.*African Journal of Science and Technology*3: 126-131p.

12. Jelfrey, A.Gwirts and Maria Nieves Garcia-Casal (2011)."Processing maize flour and corn meal food product." Department of Grain Science and Industry Kansa State university Kansa. *J. Food Science and technology.* 28:39-45

13. Kulkarni, k. D, kulkarni, D.N and Ingle, U. M. (1991): "Sorghummalt-based weaning formulation, preparation, functional properties and nutritive Values". Food and nutrition bulletin 13(4): 322-327.

14. Kurup G. T. (1994).Tuber crop starches central Tuber crops Research Institute Sreekariyam. Thiruvananihapuram, Kerala India. Tech Bulletin series 18.

15. Larmond, E., (1977): Laboratory methods for sensory Evaluation of food, Canda Pept Agric-Pub.

16. Mahgoub, S.E.O., (1999). "Production and evaluation of hydrolysate from an oilseed flour mixture. Food weaning foods based on sorghum and legumes" Chem., 106: 1166-1174. *Plants Foods Hum. Nutr.* 54: 29-42.

17. Muhammed Madeem, Salim-ur-Rahman, Faqir Muhammad Anjum, MianAnjumMurtaza and Ghulam Mueen-ud-Din(2012). "Development characterisation and optimization of protein level in Date Bars using Response Surface Methodology". Institute food Sc & Nut. University of Sargodha, Sargodha, Pakistan *J. food science 50:19-24.*

18. Narayana, K and Narasinga, Rao M. S (1984): Effect of partial proteolysis on the functional properties of winged pea (Psophocapustetragonolobus) flour. *Journal of Food Science,* 49:944-947.

19. Niba LL, Bokanga MM, Jackson FL, Schlimme D, Li BW (2001) Physicochemical properties and starch granular characteristics of flour from various manihot esculeata genotypes. *J. Food Sci.* 67:51-56.

20. Paniappan G. C (2002). Chemical compositions, antioxidant capacities and antiproliferative activities of coconut flours. *J. Agric. Food. Chem.*54:3773-3778.

21. Richard; J.R, Asaoka M.A, Blanshard J.M.V (1991). The physico-chemical properties of cassava starch. *Trop. Sci.*31:189207.

22. Safo-Kantanka K.O, Acquistucci R (1996). The physico-chemical properties of cassava starch in relation to the texture of cooked root. *Ghana J. Agric. Sci.* 28(29):69-90.

23. Steel, R.G.D and Torrie, J.H (1980): Principles and procedures of statistics. MC Graw Hill Pub. Comp. Inc. New York UNE (Una Norma Espanola).(1974). 34-074-74.

24. Sosulski, F. N. (1962): "The centrifugal method for determining flour absorptivity chemistry 39:344-346. State University Ames, Iowa.

25. Takashi, S. and Sieb, P. A. (1988): Paste and gel properties of prime corn and wheat starches with and without native Lipids". *Cereal chemistry*65:474-475.

26. Tizazu H., Emire S.A (2010). "Chemical composition, physicochemical and functional properties of Lupin(Lupinusalbus) seed grown in Ethiopia".*African Journal of Food, Agriculture, Nutrition and Development.*10(8).Pp 3029-3046.

27. Trimidad PT., H.Y Divinagracia, C.M. Aida, C.A Faarida, S.M Angelica, T.C Modesta, C.A. Askali, A.S Loyola and D.B Masa (2001) "Coconut

flour from residue: A good source of Dietary fibre". *Indian coconut journal Nutr.* 26:430-435.

28. Watson S.A (1987): Structures & Composition pp 33 – 80. In Watson S.A and Ramstad R.E (Ed) Corn Chemistry and Technology Am Soc. *Cereal chem.*68:372-378.

29. Williams, P. C. Kuzina, F. D. and Hlynka, I. A. (1970): "A rapid calorimetric procedure for estimating he amylase content of starches and starch".*Cereal Chemistry* 47(4):411-413.

30. Zhao Z. Y, K Glassman, V. Sewalt, N Wang, M. Miller, S. Chang, T. Thompson and Y. Jung (2003). Nutritionally improved transgenic sorghum. In: Vasil I. K. (ed) Plant biotechnology 2002 and beyond. Kluwer Academic Publishers, Dordrecht, pp 413.

2

PROTEIN EVALUATION OF FOODS

Malomo Olu*, Alamu Ezekiel Adediran

Department of Food Science and Nutrition, Bells University of Technology, OTA, Ogun State, Nigeria

ABSTRACT

The measurement of protein quality is usually non-specific, that is, if one compares this with the assay for minerals, one would find that in the case of minerals, a specific substance is being measured. However, with proteins, the quality depends on the quantity of at least 9 amino acids. The quality of the protein depends on the balance among the respective amino acids, on which the utilisation of the protein depends. As a result, the different methods of protein evaluation give different results for the nutritive value.

KEYWORDS

Protein Evaluation, Biological Value, Essential Amino Acids Index, Protein Efficiency Ratio, Net Protein Retention, Dinitro-Benzene

1. BIOASSAY OF PROTEIN QUALITY

The biological evaluation of protein quality is usually carried out using rats.

In 1919, Osborne et al. (1919) introduced the term Protein Efficiency Ratio (PER), a biological method based on the expression of growth-promoting value of protein numerically. This hypothesis proposed by Osborne and his co-workers was the "optimum of the protein which is determined not only by the absolute amount furnished, but also by its quality". From this work, it was observed that protein sources could be evaluated by simply comparing the relative weight gain of an animal fed the material in relation to the actual amount of protein consumed. According to Bender (1956) this was considered the simplest and most convenient method of measuring the nutritive value of proteins.

The protein efficiency ratio has been frequently criticized and Mitchell (1924) stated that simplicity was its only recommendation. There are two main drawbacks to the method. Firstly, protein efficiency ratio as measured on any protein increases with the quantity consumed (Mitchell, 1924; Barnes et al., 1945). It should be noted that a certain amount of protein is required for maintenance of body weight, and only the increment above this amount is available for growth. Consequently, when larger quantities are consumed, more is available for growth and a higher protein efficiency ratio (PER) results. Secondly, the method is based on the assumption that the increase in the body weight on the protein containing diet is proportional to the protein retained. It has been repeatedly shown that the composition of the weight increases varies with the type of diet during 6-8 weeks of the usual experiments (Hamilton, 1939).

The protein efficiency ratio is open to other criticisms. Bender and Doell (1957) pointed out that gain in body weight is constant in composition is not valid. They argued further that the result may vary with the level of protein in the diet, and that the results are influenced by food intake. Moreover, no allowance is made for maintenance; it is assumed that all the protein consumed is used for growth. The method ascribes a value of zero to protein that do not that permit growth. To correct the dependence of protein efficiency ratio on food intake Bender and Doell

(1957) proposed the use of a slope ratio method whereby weight loss of a control group of animals given a protein free diet as well as the weight gain of the test group are considered. This slope-ratio method was put forward in an attempt to overcome some of the problems in the protein efficiency ratio. It requires the testing of protein at several levels in an effort to give an estimate of the precision and validity of assay. However weight loss of the rat fed a non protein diet id equivalent to the protein needed for maintenance. In an argument put forward by Mclaughlan and Keith (1975) to criticize the slope ratio method of Bender and Doell (1957), they assumed a straight line relationship involving the weight loss of the non-protein group of rats proposed by Bender and Doell (1957), and other levels of dietary protein. The results of the slope-ratio method, otherwise known as the net protein retention value (NPR), at any point of the curve, remained constant, while on the other hand the protein efficiency ratio varied along the curve. However, this straight line relationship is an ideal situation. It is not the usual form of the response curve. Typically, the response curve is not straight but deflects downwards, cutting the y-axis at a point less negative that the weight loss of the non-protein group. Hence the net protein ratio values tend to fall with increasing protein intake.

Another modified assay also known as the slope ratio method, which is an extension of the previous method of Bender and Doell (1957), was put forward by Hegsted and Chang (1957) to avoid problems resulting from non-linear response. This slope-ratio assay is a multiple dose (protein level) method, which they termed relative nutritive value (RNV). Here three or four levels of the test protein are fed, but only the points falling on the linear portion of the curve are used in calculating the slope of the response line. The slope of the response line for the test protein is expressed as a percentage of the slope of a reference protein. Mclaughlan (1972) reported that the slope ratio method gave invalid assays for lysine-deficient proteins. Moreover, the inclusion of the non-protein group in the calculation of the slope of lysine-deficient protein group seriously skews the slope. A more complex method of evaluating protein in diets is the measurement of Biological Value (BV). This is the currently much sued method for assessing protein quality. This is calculated from nitrogen intake and nitrogen excreted by the experimented animal. This concept was first introduced by Thomas

(1909), who measured this value in adult humans. The procedure was later refined by Mitchel (1923-24) and Mitchel et al. (1945), who used growing rat and measured the requirements for both maintenance and growth. In the utilisation of nitrogen, mixtures of amino acids are involved. When the individual amino acids are absorbed, the excess over the amount needed may be deaminated and the nitrogen component of the amino acids excreted in the urine. The calculation of biological value, using the Thomas-Mitchel expression, is as follows:

$$BV = \frac{NI - (FN - FMN) - (UN - UNe)}{NI - (FN - FMN)} \times 100$$

NI = Total nitrogen iontake

FN = Total faecal nitrogen

FMN = Metabolic faecal nitrogen

UN = Total urinary nitrogen

UNe = Endogenous urinary nitrogen

The calculation of biological value (BV) by the Thomas-Mitchell method includes the correction of the total faecal nitrogen for metabolic faecal nitrogen. The urinary nitrogen excreted is also corrected for the endogenous nitrogen. But the cutaneous losses of nitrogen was ignored.

There are two serious limitations to the use of the Thomas-Mitchell expression as a practical method:

1. It is difficult to measure the endogenous urinary nitrogen.

2. The level of dietary protein does not alter the result. To eliminate this, the level of 100 percent in the diet is normally used as standard (Hamilton et al., 1952). As a practical measure, the apparent BV is sometimes measured according to the following formula:

$$\text{Apparent BV} = \frac{NI - (FN - UN)}{NI} \times 100$$

NI = Total nitrogen intake

FN = Total faecal nitrogen

UN = Urinary nitrogen

The biological nitrogen value method is time consuming and cumbersome. Bender and Miller (1953) introduced the term net protein value (NPV), but Miller and Bender (1955) later referred to the method of the nitrogen retained and that of the nitrogen intake, multiplied by 100. It is numerically the product of the biological value and digestibility.

$$(NPU) = \frac{\textit{Nitrogen retained}}{\textit{Nitrogen intake}} \times 100 = \text{Biological value} \times 100$$

They determined NPU with a group of rats after a period of ten days on test protein and nitrogen and nitrogen-free diet which provide the endogenous nitrogen, while nitrogen deposited as new tissue by animals on test diets was determined with reference to the total nitrogen consumed. The NPU therefore estimates the function of the food nitrogen that is retained by the rats. Mclaughan and Keith (1975) described rat bio-assays for protein quality to be essentially a measure of the amount of the limiting amino acid available to the test animal. They argued that since the limiting amino acid differs from protein to protein, it is possible to be measuring one amino acid in one protein, and a different amino acid in another protein. They further stressed that matters could be complicated if there to be difference in the limiting amino acid, depending upon the level of protein in the diet.

In the final analysis, one thing that is certain is that no rat bio-assay method for evalasting protein quality will ever be completely satisfactory as a predictor of nutritional value for the human diet.

2. CHEMICAL METHODS

Many procedure using chemical composition have been applied to evaluation of protein. Mitchell and Block (1946) devised a system of "chemical scores" based on the amount of the essential amino acids in a protein compared to the level present in a reference protein selected for its nutritional excellence. Mitchell (1954a) found chemical scores to have a high degree of correlation (r = 0.95) with published biological values of food proteins. However, Mitchell (1954a and b) pointed out in his works that biological values of foods depends on the content of essential amino acids, and that chemical score is an index of the value of protein for growth only, because he assumed that the absence of an essential amino acid renders the protein completely unavailable even for tissue maintenance. This assumption was not consistent with observations reported by Mason and Palmer (1935).

Shelling (1975) argued that the chemical score procedure simply involves expressing the amount of each indispensable amino acid in the test protein as a percent of that in whole egg. A lower chemical score, he further stated thereby represents a lesser amount of that amino acid relative to the amount required. However, Rao et al. (1964) concluded that the establishment of reliable rat amino acid requirements has led to their as a reference for chemical score, calculated according to Mitchell and Block (1946), using the requirement pattern as the standard, and the biological values of 15 proteins to be highly significant (r =+0.89). This they found to be compare well with that (r = +0.846) reported by Mitchell (1954a) using egg protein as standard.

Oser (1959), in his contribution to protein evaluation using amino acid analysis, stated that each amino acid may be considered as a keystone in an arch, without which a stable protein structure cannot be built. Accordingly, the rating of a protein for nutritional quality should take into account its entire contribution of essential amino acids, not merely the one in greatest deficit with respect to a nutritional standard. He therefore devised a means of representing the entire essential amino acid spectrum of a protein by an integer termed the essential amino acid index (EAAI). The essential amino acid index derivation was based on two assumptions, that the minimum ratio of essential amino acid content relative to that of standard protein is one percent, and that the maximum is 100 percent. Oser (1959) defined this mathematical model as the geometric mean of "the egg ratio", that is, the ratios of the essential amino acids in a protein relative to their respective amounts in whole egg protein:

$$\text{EAA Index} = n\sqrt{\frac{lysine\ P}{lysine\ S} \times \frac{tryptophan\ P}{tryptophan\ S} \times \frac{histidine\ P}{histidine\ S}}$$

Subscript p = food protein

S = standard protein (whole egg)

N = number of amino acids (counting pairs such as menthionine and cystine as one)

In this same work work, Oser (1959) suggested calculating the egg ratio for each individual amino acid and using these values in the computation of the essential amino acid indexes. He further suggested that, on occasions when certain amino acid values are not available, they could be omitted, provided that the appropriate 'n' is not used.

A high degree of correlation between essential amino acid indexes (EAAI) and biological values was found. Form the EAA indexes, he derived an equation for calculating the biological value (Oser, 1959):

Biological value = 1.09 (EAAI) - 11.7

The correlation between the EAA index and the biological value was so consistently high that Oser finally concluded that estimates of the latter may be from essential amino acid assay with a degree of reliability greater, in the long run, than is obtainable in the usual biological assay using 10 rats per group. However, whether the results give useful information depends on the availability to the animal organism of the determined amino acids. He found that factors such as heat treatment or insolubility, which antagonise the degree or rate of digestibility and absorption of protein, do cause complication in the interpretation of results.

Rao et al. (1964) argued that essential amino acid requirements should serve as a standard for evaluating the quality of protein in foods, rather than the total essential amino acids content of the food. This led to the development of a mathematical model based on the calculation of the geometrical mean of essential amino acids, each expressed as a percentage of its minimal requirements, except that values exceeding 100 percent are reduced to 100 percent. This mathematical model they called protein requirement index.

In this method, the minimal requirements for the essential amino acids and the optimal protein level for maximal growth and body nitrogen retention of the growing rat was used.

Kapour and Heiner (1982), using the essential amino acid index (EAAI), calculated biological value and requirement index to evaluate changes in developing wheat adn found a high correlation (r = +0.884) between the calculated biological value method of Oser (1959) and the requirement index method of Rao et al. (1964).

According to Eggum (1968), chemical assay methods are of limited value for the analysis of protein quality because digestibility and biological digestibility of the proteins are overlooked. However, Said and Hegsted (1969, 1970) found that net protein utilisation and biological value using animal studies are unsatisfactory methods for assessing the protein value of grains for humans. Thus the chemical methods used in the present study are quite useful supplements to the biological methods, and are particularly valuable in predicting amino acid deficiency.

3. AMINO ACIDS

Of the 22 amino acids known to be physiologically important, the body is capable of synthesizing some under proper conditions and if a supply of nitrogen is made available. These groups of amino are known as the dispensable, or non-essential amino acids. The rest cannot be synthesized by the body and must be therefore be supplied by diet.

These are indispensable or essential amino acids: leucine, isoleucine, lysine, phenylalanine, methionine, tryptophan, valine and threonine. Histidine may be added to theses as it appears to be essential to the growth of infants. When arginine is added to the latter group to make a total of 10, they are considered essential for the growth rat and chick. The previously mentioned 8 are essential for human adults (Rose et al., 1949; Mitchell, 1959).

Amino acid analysis, using ion exchange chromatography, has been favoured in this. Walker (1981) regarded it as one of the most important chemical techniques for predicting protein quality. However, since the chemical score obtained from amino acid analysis data may over-estimate protein quality, values for digestibility and amino acid availability are now incorporated in the score. An additional problem is that the acid hydrolysis, which is necessary for sample preparation, destroys tryptophan, and may give low values for sulphur containing amino acids. Bender (1961) had found that, despite the limitations, chemical score based on amino acid analysis is capable of judging protein quality accurately when compared with rat assay, provided the analysis is carried out carefully.

One other factor that can affect the validity of amino acid data is the way they are calculated. Blackburn (1978) suggested the addition of an internal standard of norleucine to the amino acid hydrolysates in order to make corrections for recovery losses.

3.1 Amino Acid Availability

The nutritional quality of the proteins of most foodstuffs is dependent largely on two essential amino acids, lysine and tryptophan. However, methionine, threonine and isoleucine may also be quite low in several foodstuffs. These amino acids are frequently referred to as the critical amino acids, the most deficient in relation to the standard protein being called the first-limiting amino acid (Concon, 1975).

Chemical methods for determination of amino acid availability are almost entirely restricted to the estimation of available lysine. Most of the methods are based on the estimation of free ϵ - amino group of lysine in the intact protein. The nutritional availability of lysine has been based on the assumption that these groups are free to react chemically. Lysine in its reactive (available) form is reported to be the first-limiting essential amino acid to be rendered unavailable during processing, or even prolonged storage under certain conditions (Carpenter, 1974; Concon, 1975). Thus lysine may easily deaminate, be exidised, or undergo the browning reaction. Therefore its availability can be used to monitor processing damage, and other unit operations.

The most widely used chemical test for the estimation of the lysine is based on the reaction of the ϵ-amino group of lysine with 1-fluoro-2, 4-dinitro benzene (FDNB).

A procedure for the estimation in foods of the lysine units whose ϵ-amino -amino group will undergo reaction with fluorodinitrobenzene (FDNB), as illustrated in figure 3, was developed by Carpenter and Elinger (1955). Any lysine residue which is already bound with another group through its ϵ-amino group is not free, which is available for reaction with the reagent, produce a yellow dinitrophenyl derivative (DNP), which can be estimated colorimetrically after acid hydrolysis.

According to Bruno and Carpenter (1957), α-DNP-arginine-,γ-DNP ornithine and water-soluble breakdown products from the other DNP – amino acids still remained in the aqueous phase along with the ϵ-DNP-lysine. A modification was introduced by these workers to correct for these error. The colour contributed by theses compounds was estimated after removal of the ϵ-DNP-lysine from the aqueous phase. Reaction of the yellow aqueous phase with methoxy carbonyl chloride changed the ϵ-DPN-lysine into another soluble derivative, the interfering compounds remaining water soluble. Colorimetric measurement of the yellow aqueous phase before and after treatment with methoxycarbonyl chloride and other extraction accordingly gives an estimate of the colour contributed by the impurities.

The correction for the loss of ϵ-DNP lysine produced during hydrolysis was carried out by Carpenter (1960). He measured the recovery of ϵ-

DNP lysine added to the DNP –protein immediately before the hydrolysis stage. An average value of 92 percent was found. He then ccorrected his available lysine value on this basis, with the assumption that free ∈-DNP-lysine similarly on hydrolysis to ∈-DNP –lysine in proteins.

Booth (1971) found that added ∈-DNP- lysine was partly destroyed during acid hydrolysis, but protein bound DNP-lysine was destroyed to a lesser extent. The differences in destruction of added $\in -DNP$ lysine was considered to be due to the DNP –protein being less soluble than ∈-DNP-lysine and so less readily susceptible to reduction by breakdown products of carbohydrates. The latter are produced in early stages of acid hydrolysis, where ∈-DNP lysine is only slowly released from DNP –protein. The breakdown products of carbohydrates quickly react with soluble nitrophenyl compounds and are mostly removed by the time that an appreciable amount of ∈-DNP- lysine has been released from the protein.

Carpenter's method is well established from measuring available lysine in animal protein products; difficulties are encountered when it is applied to cereal grains and legumes. These are manifested as poor recoveries of ∈-DNP-lysine and variable results. The reasons for the poorer performance of the method on these materials are associated with the destruction of ∈-DNP-lysine by carbohydrate during hydrolysis and the formation of other yellow products which are not easily separated from ∈-DNP-lysine (Bruno and Carpenter, 1957; Rao et al., 1963). Rao et al (1963), using Carpenter's hydrolysis technique followed by ion exchange chromatography to isolate, ∈-DNP-lysine formed, found that a good recovery adn reproducibility with cotton seed meals. Roach et al (1967) measured available lysine in high protein animal and vegetable meals using a semi-automatic method based on the measurement of total lysine and the lysine remaining after treatment of the meals with fluorodinitrobenzene according to the Carpenter method. The difference between these two values represents the lysine in the protein which has free ∈-amino group and is expressed as available lysine. Their results compared well with those of the established method of Carpenter (1960) and Rao et al. (1963).

Carpenter (1960) recommended that a correction factor of 1.09 should be used for animal and fish proteins to compensate for and average recovery of 92 percent. Roach et al. (1967), on the other hand, made individual corrections based on the actual recoveries samples, and that the variability of the recoveries increased the standard deviation over that of the uncorrected results. This evidence therefore makes the use of an average correction factor recommended by Carpenter (1960) questionable.

Walker (1979b), using the Carpenter (1960) method, as modified by Booth (1971), found that the FDNB-reactive lysine was highly correlated (r =+0.90) with the dye-binding reactive-lysine method. Dye-binding, using acid dyes at low pH, primarily involves the formation of ionic linkages with the basic amino acids (lysine, histidine and arginine) present in protein, as well as with terminal amino groups (Walker, 1979a).

Others have attempted to separate DNP-lysine from interfering amino acids and other compounds formed during hydrolysis of carbohydrate containing food by paper chromatography (Balinga et al., 1959), thin layer chromatography (Holm, 1971), or liquid-liquid partition chromatography (Blom et al., 1967).

Peterson and Wartesen (1979) also attempted to use high pressure liquid chromatography (HPLC) to determine available lysine using a modification of Carpenter's (1960) method.

The laboratory evaluation of protein quality in terms of lysine availability using the Carpenter (1960) chemical method is found relatively simple and accurate. Hence the use has been favoured in the present study for protein quality measurement.

4. IN-VITRO DIGESTIBILITY

Chemical score and amino acid index procedures, widely used for evaluating the protein quality of foodstuffs, have been earlier reviewed. Although these procedures are rapid, and in many cases accurate;

however, no allowance was made for variations in digestibility and availability of the amino acids. Various methods have been proposed for the in-vitro evaluation of digestibility.

The enzyme system developed by Sheffner et al. (1956) involved determining the essential amino acids released by in-vitro pepsin digestion and those in the remainder of the protein to calculate a "pepsin-digest residue (PDR) amino acid index.

Saunders et al. (1973), in their in-vitro digestibility of protein studies using pepsin, pepsin-pancreatin and pepsin trypsin methods, found that values obtained with either system involving pepsin showed a high degree of correlation with protein digestibility values obtained in rat feeding trails.

Hsu et al. (1977) developed a rapid multi-enzyme technique for estimating protein digestibility. The multi-enzyme system consisted of trypsin, chymotrypsin and peptidase. They found that the pH of a protein in suspension immediately after digestion for 10 minutes with the multi-enzyme solution was highly correlated with the in-vivo apparent digestibility in rats. Regression analyses of 23 samples tested by these workers showed that the correlation coefficient between pH at 10 minutes and in-vivo apparent digestibility was 0.90 with a standard error estimate of 2.23. The regression equation obtained was $Y = 210.464 - 18.103\ X$, where x was the pH of protein suspension immediately after 10 minutes digestion with the multi-enzyme solution. They claimed that the most significant advantage of this in-vitro method for prediction apparent protein digestibility over others was that it would be completed within one hour and with a high degree of sensitivity. They also claimed that the method could detect the effects of trypsin inhibitor, chlorogenic acid and heat treatment on protein digestibility. Although strong buffer salts may affect the measurement of protein digestibility, the buffering effects found in general food proteins and product tested did not create any problem with this procedure.

Walker (1981) found a high correlation (r = +089) between in-vitro digestibility using the Saunders et al. (1973) method, and available lysine by dye-binding with 7 various leaf protein concentrates. The protein digestibility of Saunders et al. (1973) was used in the present study, because of its high correlation with bio-assay methods and other chemical methods of investigating protein quality, as reported by Walker (1981).

5. ENERGY – PROTEIN RELATIONSHIPS

The estimates of protein requirements are valid only when energy requirements are fully met. When the total energy intake is inadequate, some dietary protein is used for energy and is not available to satisfy protein needs. The further increase in protein intakes to meet safe levels is of limited efficiency and wasteful, if energy needs are not met at the same time (FAO/WHO, 1973).

The maximum intake of dietary protein necessary to achieve nitrogen balance depends not only on the amino acid composition and the digestibility of the protein, but also on the composition and adequacy of the diet as a whole (Young and Scrimshaw, 1978). Among the most important dietary factors that affect dietary nitrogen retention are:

i. The quantity of protein and non-protein energy in the diet

ii. The amount of total nitrogen

iii. The intake of essential amino acids.

The interactions between dietary protein and energy intake have been the subject of many studies in humans as well as in farm and laboratory animals (Richardson et al., 1975).

Studies with humans (Young and Scrimshaw, 1978; Wilmore, 1977) indicate that protein metabolism is highly sensitive to energy intake, the level of energy intake influencing the estimated minimum requirement for dietary protein, and the effects of energy and protein intakes have important implications for quantitative estimation of human protein amino acid requirements (Garza et al., 1973).

Calloway (1980) found that, in both children and adults, male and female, increasing energy of the diet from marginally sub-maintenance levels to luxus levels resulted in a more positive or less negative balance. It was argued that this effect requires the presence of sufficient protein of adequate quality to allow N balance.

This study investigated changes in energy-protein ratio in germinating grains and the effect of processing on this ratio.

REFERENCES

1. AKESON, W. R., AND STAHMAN M. A. (1964). A pepsin pancreatin digest index of protein qualityevaluation. J. Nutri. 83; 257.
2. BALINGA, B. P., BAYLISS M. E. And LYMAN C. M. (1959). Determination of free lysine-amino groups in cotton seed meals and preliminary studies on relation to protein quality. Arch. Biochem. Biophys. 84: 1.
3. BENDER, A. E. And MILLER, D.S. (1953). A new brief method of estimating rat protein value. Biochem. J. 53, VII-VII
4. BENDER, A. E. (1971). Processing damage to protein foods. Paper presented at 18th PAG. Meeting Room (1971), PAG BULL., p. 10.
5. BENDER, A. E. And DOELL, B. H. (1957). Biological evaluation of protein; a new aspect. Brit. J. Nutri. 11: 140
6. BENDER, A. E. (1961). Determination of nutritive value of protein by chemical analysis. In: Meeting protein needs of infants and children pub. 843. National Academy of Sciences. Nat. Research Council, Washing, D. C. Pp 407.
7. BLOM., HENDRICKS, P. and CARLS, J. (1967). Determination of available lysine in foods. Anal. Biochem. 21: 382.
8. BOOTH, V. H. (1971). Problems in the determination FDNB available lysine. J. Sci. Food Sci. 40: 266

9. BRUNO AND CARPENTER, R. J. (1957). A modified procedure for the estimation of available lysine in protein. Biochem. J. 67: 13.

10. CARPENTER, K. J. (1960). The estimation of available lysine in animal protein foods. Biochem. J. 77: 604.

11. CARPENTER, K. J. (1974). Chemical and Microbial assays for the evaluation of processed protein foods. In: Nutrients in processed food proteins. American Medical Association (eds. P. L. White and D. C. Fletcher). Action, Massachussetts. Pub. Sciences Group Inc.

12. CALLOWAY, D. H. (1980). Energy-protein relations. Unpublished. Univerty of Carlifornia, Berkley. Dept. Of Nutritional Sciences, Berkley Ca. (Obtained from Dr. Richardson, Lord Zucherman Laboratory, University of Reading, Reading)

13. CONCON, J. M. (1975). Chemical determination of critical amino acids in cereal grainsa dn other foodstuffs. In: Protein nutritional quality of foods and feeds. Part 1 (ed. M. Friedman). Marcel Dekker Inc., New York.

14. EGGUM, B. D. (1968). Evaluation of protein quality and the development of screening techniques. Panel on new approaches to breeding for plant protein improvement held at Svalof, Sweden, 17-21 June 1968.

15. GARZA C., et al (1973). Human protein requirements – evaluation of the 1973 FOA/WHO safe level of protein intake for young men at high energy intakes. Brit. J. Nutr. 37: 403

16. HAMILTON, T. S. (1939). The growth, activity and composition of arts fed diets balanced and unbalanced with respect to protein. J. Nutr. 17: 565.

17. HS,U H. W., VARAK, D. L., SATHERLEE, L. D. And MILNER, G. A. (1977). A multi enzyme technique for estimating protein digestibility. J. Food. Sci. 42: 42: 1269.

18. HEGSTED, D. M. and CHANG Y. O. (1965). Protein utilisation in growing rats. 1. Relative growth index as a bioassay procedure. J. Nutr. 85: 159.

19. HOL M, H. (1971). Micro-method for determination of non-N-terminal lysine in milk and similar products. J. Sci. Food Agric. 22: 378.

20. KAPOUR, A. C. And HEINER R. E. (1982). Biochemical changes in developing wheat grains. Changes in nitrogen fractions, amino acids and nutritional quality. J. Sci. Food Agric. 33: 35.

21. MASON, I. D. and PALMER L. (1935). Utilisation of gelatine, casein and zein by adult rats. J. Nutri. 9: 489.

22. McLANGHLAN, J. M. (1972). Nutritional evaluation of proteins by biological methods. Cereal Sci. Today 117: 162.

23. McLANGHLAN, J. M. and KEITH M. O. (1975). Bioassays for protein quality. In: protein nutritional quality of foods and feeds. Vol. I (ed. M. Friedman). Marcel Dekker Inc., New York.

24. MILLE,R D. S. and BENDER, A. E. (1955). The determination of the net utilisation of proteins by a shortened method. Brit. J. Nutri. 9: 382.

25. MALOMO, O. (1982). Protein evaluation of foods. In. Germination as a unit: Its application in the development of a cereal-legume food product. PhD Thesis, University of Reading, UK.

26. MITCHELL, H. H. And BLOCK, R. J. (1946). Some relationship between the amino acid content of protein and their nutritive value for the rat. J. Biol. Chem. 163: 599.

27. OSBOURNE, T. B. And MENDEL, L. B. (1919a). Nutritive properties of proteins of maize kernel. J. Biol. Chem.. 18: 1.

28. OSER, B. L. (1959). An integrated essential amino acid index for predicting the Biological value of proteins. In: Protein and amino

acid nutrition (ed. A. A. Albanese). Academic Press, New York PP 281.

29. RAO, S. R., CARTER F. C., and FRAMPTON V. L. (1963). Determination of available lysine in oil seed meal proteins. Anal. Chem.. 35: 1927.

30. RAO, P. B. R., NORTON H. W. And JOHNSN, B. C. (1964). The amino acid composition and nutritive value of proteins. V. The amino acid requirements as a pattern for protein evaluation. J. Nutr. 82: 88.

31. RICHARDSON et al (1975). Quantitative effect of an iso-energetic exchange of fat for carbohydrate on dietary protein utilisation in healty young men. Amer. J. Clinical Nutr. 32: 2217.

32. ROACH, A. G., SANDERSON, P. And WILLIAMS, D. R. (1967). Comparison of methods for the determination of available lysine values in animal adn vegetable protein sources. J. Sci. Food Agric. 18: 274.

33. SAID, A. K. and HEGSTED, D. M. (1969). Evaluation of dietary protein quality in adult rats. J. Nutr. 99: 474.

34. SAID, A. K. and HEGSTED, D. M. (1970). Response of adult rats to low dietary levels of essential amino acids. J, Nutr. 100: 1363.

35. SAUNDERS et al (1973). Measurement of digestibility of alfalfa protein concentrates by vitro methods j. Nutr. 103: 530.

36. SCHELLING, G. T. (1975). An efficient procedure for the complet evaluation of dietary proteins. In: Protein Nutritional quality of foods and feeds. Vol. 1 (ed. M. Friedman). Marcel Dekker Inc., New York. P. 137.

37. SHEFFENER, A. L., ECKFELDT, G. A. and SPECTOR, H. (1956). The pepsin digest residue (PDR) amino acid index of net protein utilisation. J. Nutr. 60: 105.

38. WALKER, A. F. (1979a). Determination of protein and reactive lysine in leaf protein concentrate by dye binding. Br. J. Nutr. 42: 445.

39. WALKER, A. F. (1981). The application of in-vitro methods to the estimation of protein quality of leaf protein concentrates. PhD Thesis, University of Reading, England.

40. YOUNG, V. R. and SCRIMSHAW, M. S. (1978). Nutritional Evaluation of proteins and protein requirements. In: Protein Resources and Technology (ed. M. Milner, M. S. Scrimshaw and D.I. C. Wang). Westport, Connecticut. AVI Publishoing Co., p. 136.

3

INCORPORATION OF COCONUT FLOUR IN PLAIN CAKE AND INVESTIGATION OF THE EFFECT OF SUGAR AND BAKING POWDER ON ITS BAKING QUALITY

Shakhawat Hossain[1], Mohammad Rezaul Islam Shishir[2], Md. Saifullah[1, 2, *], Md. Shahidullah Kayshar[3], Samshad Wasit Tonmoy[1], Afzal Rahman[1], Md. Shams-Ud-Din[1]

[1]Department of Food Technology and Rural Industries, Faculty of Agricultural Engineering and Technology, Bangladesh Agricultural University, Mymensingh, Bangladesh

[2]Department of Process and Food Engineering, Faculty of Engineering, University Putra Malaysia, Serdang, Selangor, Malaysia

[3]Department of Food Engineering and Technology, Faculty of Agricultural Engineering and Technology, Sylhet Agricultural University, Sylhet, Bangladesh

ABSTRACT

Coconut flour is an excellent source of unique taste and aroma and rich in vitamins, minerals and dietary fibers, which might have potential application in baking products and human nutrition. The study is aimed to investigate the effect of sugar and baking powder on the plain cake incorporated with different levels of coconut flour. The coconut fleshes were dried by mechanical dryer for 6 hr and grinded to coconut flour.

Four types of cake such as, S_1=0%, S_2=10%, S_3=20% and S_4=30% of coconut flour incorporation were investigated and S_3 is found more acceptable in terms of physicochemical properties of cake. S_3 cake secured the highest score in color, texture and overall acceptability. The sugar added (20-100%) increased the cake weight and decreased the moisture of cake, in which the cake volume and specific volume increased up to 80% sugar addition in cake dough. In contrast, the increase of baking powder (1-8%) decreased the cake weight and moisture, but increased the cake volume and specific volume up to 7% baking powder addition. However, S_3 cake revealed the maximum output and better acceptability at 80% sugar and 7% baking powder addition.

KEYWORDS

Coconut Flour, Cake, Sugar, Baking Powder, Baking Quality

1. INTRODUCTION

The coconut (*Cocos nucifera Lin.*) is the most extensively grown and used nut in the world. Coconut is a good source of energy, because it contains approximately 37.29% fat, 11.29% carbohydrate and 4.08% protein [1]. The color is pure white, the fragrance is sweet, and the eating is superb. Coconut is an important plantation crop of Bangladesh and called the nurse of India. It plays an important role in the diet of the people in the world. Virgin coconut oil (VCO) is a recently emerging high demand product in the world and various types of cold presses are used for extraction of VCO from the coconut kernel at low temperature. The whitish residue remained after extracting VCO from cold press can be milled to flour. Coconut flour can provide not only value added income to the industry, but also a nutritious and healthy source of dietary fiber [2]. Coconut flour may play a role in controlling cholesterol and sugar levels in blood and prevention of colon cancer. Studies revealed that consumption of high fiber coconut flour increases fecal bulk [3].

Cake is one of the relished and palatable baked products prepared from flour, sugar, shortening, baking powder, egg, essence as principal ingredients. The variation in these constituents or formulation causes the changes in textural properties of [4]. According to Hesso et al. [5], baking process plays an important role in the structural, textural and physical properties of cakes. Cakes are one of the most popular products in bakery industry, because it is economically cheaper as well as consider being luxurious gifts for infants and school going children, who are underweight [6]. Bakery products are generally used as a source for incorporation of different nutritionally rich ingredients for their diversification [7]. Various types of nutritious cakes have been prepared by fortifying the wheat flour with coconut residue, and received popularity being nutritionally rich in protein and vitamins [8].

Nowadays, emerging beneficial oil that comes in picture of world is virgin coconut oil [9]. The major difference between coconut oil and virgin coconut oil is the method of extraction. The general common method for the preparation of coconut oil is drying and later extraction of oil from kernel in general called copra. On the other hand, virgin coconut oil is extracted from fresh coconut (not copra) Kernel or meat by mechanical or natural means under controlled temperature. Usually, meal obtained after the extraction of oil, have been found in animal use feed like poultry, fish and swine industry. However, meal obtained after oil extraction still possess good nutritional properties which could be utilized for value addition of various processed foods. Dairo, [10] has reported the protein quality indices of sun dried coconut oil meal; oven dried coconut oil meal and fermented coconut oil meal on rats. However, data on the use of Virgin coconut meal in the development of various processed foods are scanty.

The consumer of Bangladesh is seeking convenience food items in the market due to their higher income and active life style. Thus, cakes could be processed by incorporating coconut flour with wheat flour that gave better physical, chemical and nutritional properties than plain cake. Therefore, the study is aimed to investigate the effect of sugar and baking powder on the plain cake incorporated with different levels of coconut powder.

2. MATERIALS AND METHODS

2.1 Materials

Chemicals and solvent were AR grade use from the laboratory stock, Department of Food Technology and Rural Industries, Bangladesh Agricultural University (BAU), Mymensingh, Bangladesh. Coconut, egg, sugar, powder milk, baking powder, shortening (soybean oil) and other materials were procured from the local market. High density polyethylene bags were used for storage of samples.

2.2 Preparation of Coconut Flour

Coconut flour was processed from the local variety of coconuts, which were free from immature and field damage. After de-husking and splitting the coconut, water was drained. Then the kernel (meat) was collected and removed milk by crushing with hand crusher and dried by mechanical dryer for 6 hours at 60ºC. The dried coconut was grinded to produce flour. The coconut flour was packed in a high-density polyethylene, sealed and stored.

2.3 Formulation of Cakes Incorporation with Coconut Flour

The basic formulation of plain cake (multi-stage mixing) [11] and composite flour cakes are wheat flour (100 g), water (50 g), sugar (40 g), oil (40 g), baking powder (3.5 g) and egg (35 g). The replacements of wheat flour in the formulations are made with 10%, 20% and 30% coconut flour followed by [12] and recognized as S_1=control cake (without coconut flour), S_2=cake containing 10% coconut flour, S_3=cake containing 20% coconut flour and S_4=cake containing 30% coconut flour. Sugar was also added at the rate of 20 to 100% and baking powder was added at the rate of 1.0 to 8.0% on the wheat-coconut flour weight basis. The wheat flour, coconut flour and other ingredients for each cake were weighed accurately and the sugar and shortening were mixed in a mixing machine for 20 minutes to produce a cream. In later stages, half of the water, egg and other ingredients and finally the flour were mixed using a mixer at low speed (145 rpm) for 10 minutes to ensure even distribution of the components. The bowl was scrapped and batter was mixed for an additional two minutes at medium speed (250 rpm). The remaining water was added and the batter mixed at low speed for two minutes. After the bowl was scrapped, the batter was mixed an additional two minutes at medium speed. Portion of batter weighing 150 g was scaled into pre-greased cake pan. All cakes were baked in national forced convection oven for 40 minutes at 170ºC.

2.4 Analysis of Coconut Cake Properties

Moisture content was determined adopting AOAC, [13]. About 5 g sample was taken in a pre-weighted crucible, which was previously heated to 130ºC. The sample was dried for 1 hour in an air oven 130 ± 3ºC. The loss of weight from sample was determined and calculated the moisture content. Protein, ash, fat and crude fiber were determined using AOAC, [13]. Total carbohydrate content of the sample was determined as total carbohydrate by difference, which is by subtracting the measured protein, fat, ash and moisture from 100 g [14]. The cake volume was determined by seed displacement method [15]. The weights and specific volume of baked cakes were also measured. The symmetry and the characteristics of crust and crumb of the cakes supplemented with coconut flour were evaluated and recorded. Cakes were evaluated organoleptically for color, flavor, texture and overall acceptability. A 1-9 point hedonic rating test was also performed to assess the degree of acceptability of cakes containing coconut flour. One slice from each lot of cake was presented to 10 panelists as randomly coded samples. The taste panelists were asked to rate the sample for color, flavor, texture and overall acceptability on a 1- 9 point scale, where 1=dislike extremely; 2=dislike very much; 3=dislike moderately; 4=dislike slightly; 5=neither like nor dislike; 6=like slightly; 7=like moderately; 3=like very much; 9=like extremely.

2.5 Statistical Analysis

The data obtained were analyzed and interpreted by analysis of variance (ANOVA) and Duncan's Multiple Range Test (DMRT) at a level of 5% of significance, using Statistical Analysis System (SAS 9.3 TS L1M2, SAS Institute Inc., NC, USA).

3. RESULTS AND DISCUSSION

3.1 Effect of Coconut Flour on Baking Quality

3.1.1 Chemical Composition of Plain Cake

The moisture content of cake containing coconut flour is shown in Table 1. It can be seen that the addition of coconut flour increased the moisture content in cake. The cake (S_4) containing 30% coconut flour gave the maximum moisture content around 20.5%. The high concentration of fibers may cause significant increase in water holding capacity of the coconut flour cakes. Similar observation was found on addition of different concentration of corn bran in the formulation of cake batter, in which the final moisture content increased with the increasing concentration of corn bran [16]. The moisture content of the S_1 cake (without coconut flour) under study was around 19.0%, which is higher than those found by Leung et al. [17], who reported 12.0% moisture on plain cake. This variation in moisture content might be due to difference in oven temperature and baking time of the cake. It is observed that protein content in coconut flour cake samples increased from S_2 to S_4 with increased substitution level of coconut flour in dough. Similarly, the fat, ash and crude fiber increased with the addition of coconut flour. The cake S_4 showed the maximum value of protein, fat, ash and crude fiber, while S_1 having the lowest in protein, fat, ash and crude fiber (Table 1). This is might be due to the coconut flour contain higher amount of protein, fat, ash and crude fiber than the wheat flour. However, the total carbohydrate content of different cakes varied from 30.05% to 52.82% being highest for cake S_1 and the lowest for the cake S_4. The total carbohydrate content (52.82%) of the cake S_1 was significantly lower than those reported by Leung et al. [14] who mentioned 79.2% total carbohydrate content in plain cake. The variation in the carbohydrate contents among cake samples under study due to the difference in protein, fat, ash and moisture content.

Table 1. Chemical composition of plain cakes.

Components (%)	Cake samples			
	S_1	S_2	S_3	S_4
Moisture	19	19.2	19.75	20.5
Protein	6.50	6.80	7.08	7.40
Fat	20.23	26.50	34.07	39.10
Ash	0.93	0.98	1.02	1.15
Crude fiber	0.52	1.05	1.42	1.80
Carbohydrate	52.82	45.47	36.66	30.05

3.1.2 Physical Characteristics of Plain Cake

(i).Volume, weight, specific volume and moisture characteristics

Volume is one of the most important physical characteristics of baked products and can affect consumer preferences [18]. It is a quantitative measurement and correlates well with dough handling properties, crumb, texture, freshness and technological versatility [19]. It can be seen from Fig. 1(a) that 20% coconut flour in cake formulations gave highest cake volume than those of 10% and 30% of coconut flour incorporation. This is due to presence of water absorbing matrix (oil, cellulose, hemicelluloses, lignin and other dietary fiber components) in coconut flour, which increased water holding capacity leading to enhancement of cake volume. A significant impact of fiber is to reduce dough gas retention and of fat is to increase dough gas retention [20,21]. The weights of cake prepared from coconut flour samples were higher than those of control cake (Fig 1. b). Cakes at 10.0%, 20.0% and 30.0% replacement level gave a higher weight due to increased water holding capacity of the coconut flour. This variation in cake weight may result from the increased water absorption by the coconut flour. The physical nature of the product as well as the form of water present in the material might have affected the results.

The fiber holds the water which contributes to the higher weight of the composite flour cake [22,23]. The specific volume of cake increased with the addition of coconut flour until 20% of coconut flour as like as loaf volume. The lowest specific volume was observed at S_4 cake having maximum weight (Fig. 1. c). In case of moisture property, the final moisture content increased with the addition on coconut flour, which was similar with the trend of cake weight (Fig. 1. d). The higher the weight of cake higher the moisture content of cake. This could be attributed to the presence of higher concentration of fiber components (cellulose, hemicellulose or pentosans, lignins and other dietary fiber) and oil might have significance for increased water holding capacity in the coconut flour dough.

(ii). Symmetry, bloom, crust and crumb characteristics

The term "symmetry" is self-explanatory. The most common faults for while points are deducted are:

Low edges, high edges, low centers, high centers and unevenness. From Table 2, it is seen that the control cake (S_1) and cake (S_3) with 20% coconut flour had better symmetry compared to other cakes.

The term bloom refers to luster or sheen, which describes the brilliance of the color. The control and coconut flour samples possess better bloom at different substitution level except cake S_4. Crust color differences of cakes containing coconut flour relative to the control are shown in Table 2. In general, the differences in crust color became larger as the substitution levels increased from 10% to 30%. The crust color varied from light brown to deep brown from cake S_1 to S_4. Cake S_3 showed the attractive color brown, which is more acceptable than others. It can be observed visually from the Fig. 2. It showed that, S_1 is yellow in color, S_2 and S_3 looking brown and S_4 is a mixture of yellow and brown with undesired shape. However, cake S_3 is mostly desired in crust color. This term crust consistency applies to the condition of the crust and varies with the types of cakes. As shown in Table 2, the tender crust was obtained in S_1 and S_4 cake. The cake sample S_4 had medium tough crust. The overall crust characteristics of 20% coconut flour cake (S_3) seem to be better than the other samples (S_2 and S_4).

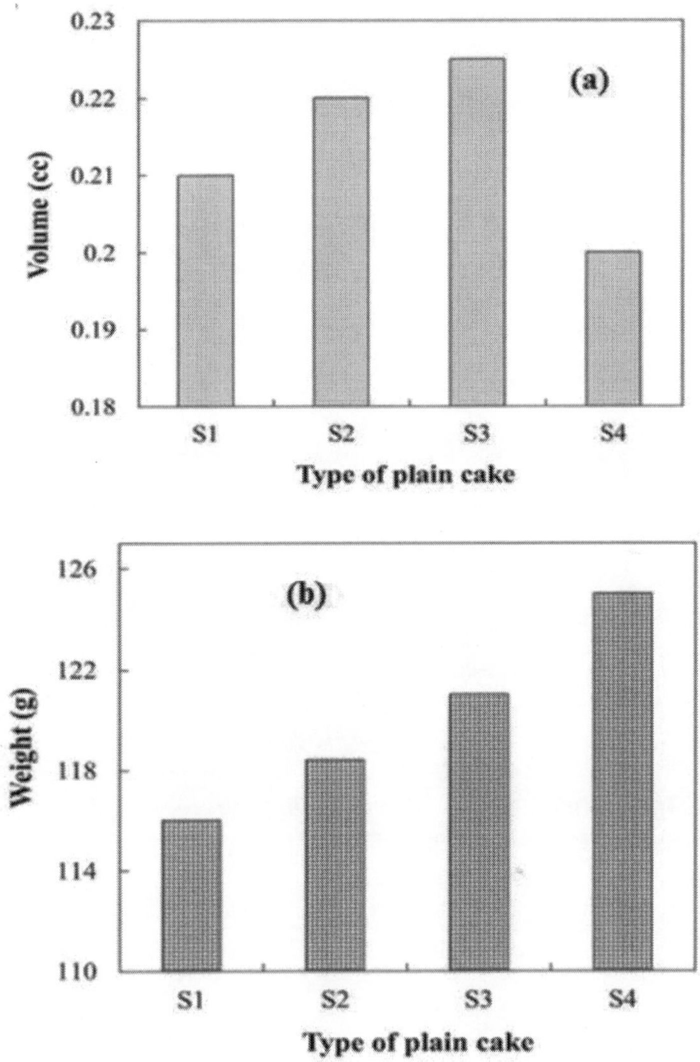

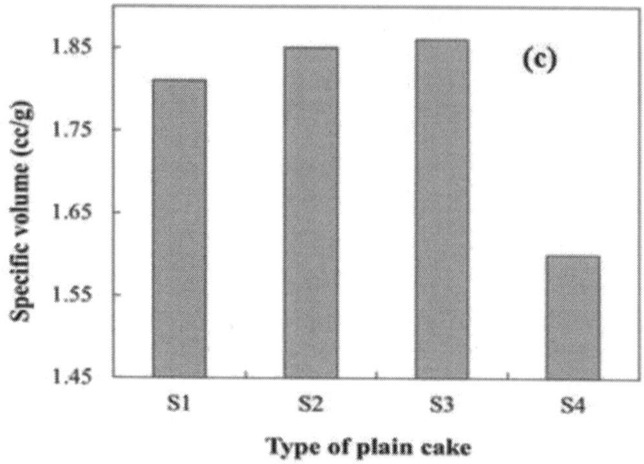

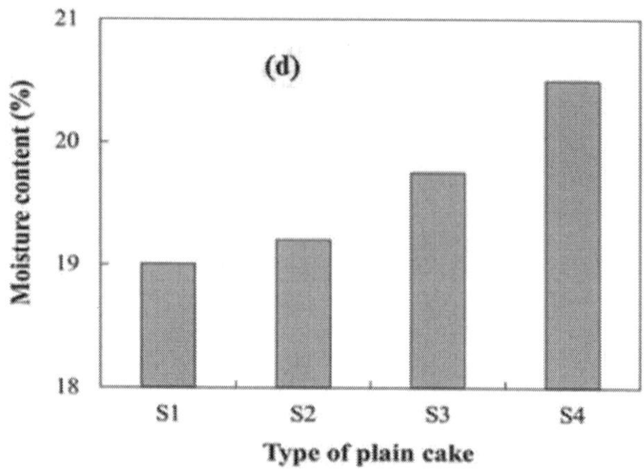

Figure. 1. Volume, weight, specific volume and moisture content characteristics of different types of cake.

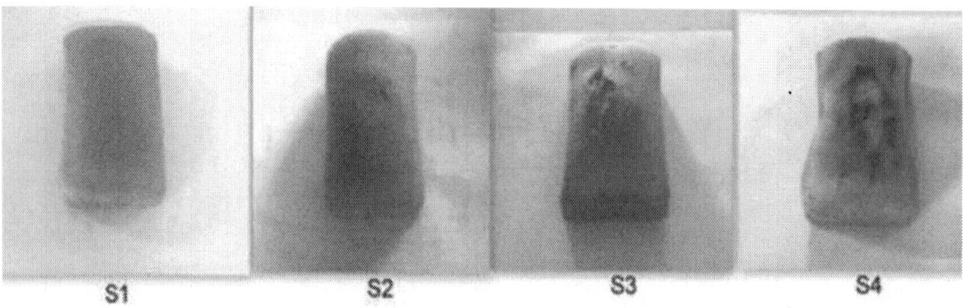

Figure. 2. Physical appearance of plain cakes.

Table 2. Symmetry, crust and crumb characteristics of plain cakes.

Cake type	Symmetry			Bloom	Crust characteristics		Crumb characteristics					
								Texture			Grain	
	Evenness	Edges	Centre		Color	Consistency	Color	Lumps and harshness	Surface	Odor	Close or airy	Shape and size
S1	Even	Medium	Medium	Luster	Light brown	Tender	White yellow	Free	Smooth, silky	Appetizing	Close	Uniform, thin walled cells
S2	Even	Low	Medium	Shining	Brownish	Medium tender	Yellowish	Free	Smooth	Slightly fresh	Close	Uniform
S3	Even	Medium	Medium	Luster	Brown	Tender	Slightly yellow	Slightly free	Light smooth	Appetizing	Close	Uniform
S4	Medium even	Too low	Low	Slightly dull	Deep brown	Medium tough	Deep yellow	Present	Rough	Sweet	Less airy	Less uniform

The crumb color of the cake supplementing by coconut flour was generally more brownish than the control cake (S_1). The increase in the levels of coconut flour substitution changed the crumb color of the coconut flour cakes from yellowish to deep gray yellow (Table 2). The crumb texture differences were observed between control and coconut flour cakes and these differences increased at the 30% substitution levels. The texture differences between control and coconut flour cakes were less perfect texture should be free from lumps and harshness and have a smooth silky surface which was obtained in the control cake (S_1) and coconut flour cake (S_2) at 10% substitution level (Table 2). The acceptable crumb odor (i.e. fresh, sweet, and natural appetizing) was found in coconut flour both at 20.0% and 30% substituted coconut cake (S_3 and S_4) and control cake (S_1). Crumb grain of cakes indicates to the shape, size and character of the cell wall structure of crumb. Uniformity of size with thin walled cell is the most desirable for crumb grain. Coarseness, thick walled cells, uneven cell size and large holes are indicative of poor grain. The cake (S4) containing coconut flour at 30% had coarser grain and the size and shape was non-uniform. For the cake (S2 and S3) containing coconut flour at 10% and 20.0%, respectively, the crumb grain was equally acceptable with the control cake (S1). This acceptability decreased for higher substitution level of 30% (Table 2).

3.2 Effect of Sugar Addition on Baking Quality

3.2.1 Volume and Specific Volume

The increasing levels of sugar significantly affected the volume of cakes. It was observed that the cake volume increased sharply up to the addition of sugar at 40% and thereafter the volume of the cake increased slowly until the addition of sugar at 80% and then fall the volume with addition of 90-100% sugar (Table 3). However, the highest volume of S_1 to S_4 were achieved (210, 220, 225, 200 cc, respectively) at 80.0% sugar level, after which the cake volume decreased progressively. At the low levels of sugar addition (20.0% or 30.0%), all the cake samples baked too early indicating less time requirement to set the cake in the oven. The top surface of the cake set before the center and expansion of the center portion was continued, which accompanied with burst in the crust. The color of the crust became pale at 20.0% or 30.0% levels of sugar addition. At high levels of sugar addition (90.0% or 100.0%) a thicker crust was formed on the cake and the lower half of the cake crumb discolored. This might be due to the increasing rate of caramelization of sugar during baking. At this level (90.0% or 100.0%) of sugar addition, most of the cakes had crispy crust, leaving sugar particles on the surface of the cakes. Although 90.0% levels of sugar addition provided highest volume in all types of cakes except control, the acceptable quality cake was obtained at 60.0% levels of sugar addition in terms of texture, cell structure, air holes, flavor, shape and size of the crumb grain, setting time and color of both crumb and crust of the cakes.

3.2.2 Weight and Moisture Content

The effect of various levels of sugar on the cake weight and moisture contents was evaluated and the results are shown in Table 4. The weight of S_1 varied significantly from 108 to 116 g the highest is for 80% sugar level and the lowest is for 20% sugar level. It is also seen that all other samples also attained in making weight at 80% sugar addition and the higher the substitution the higher is the weight. This may be due to higher percent moisture content retained by the samples having progress higher substitution level of coconut flour at sugar content. The highest moisture content (24.0%) was obtained S_1 at 20% sugar addition, while the lowest moisture content (18.0%) was found at the 90% (or 100%) sugar addition. With different levels of sugar addition from 20 to 100% the weight of all the cakes with coconut flour varied remarkably from 113 to 125 g. The highest weight of the cake sample S_4 (125 g) was achieved at 80% and 100% sugar addition and the lowest cake weight (108 g) of the sample S_2 was obtained with 20% sugar addition among all samples except control sample.

Table 3. Effect of sugar level on volume and specific volume of plain cakes.

Sugar level (%)	Cake sample							
	S_1		S_2		S_3		S_4	
	Volume (cc)	**Specific volume (cc/g)**	**Volume (cc)**	**Specific volume (cc/g)**	**Volume (cc)**	**Specific volume (cc/g)**	**Volume (cc)**	**Specific volume (cc/g)**
20	130	1.20	138	1.25	142	1.29	125	1.10
30	160	1.45	170	1.50	175	1.57	150	1.31
40	190	1.71	195	1.69	200	1.78	185	1.56
50	200	1.76	205	1.76	210	1.81	191	1.56
60	205	1.79	212	1.79	215	1.83	196	1.59
70	208	1.80	217	1.83	221	1.84	198	1.59
80	210	1.81	220	1.85	225	1.86	200	1.6
90	208	1.82	216	1.84	220	1.84	197	1.6
100	200	1.76	214	1.82	217	1.85	190	1.57

Table 4. Effect sugar level on weight and moisture content of plain cakes.

Sugar level (%)	Cake samples							
	S_1		S_2		S_3		S_4	
	Weight (g)	Moisture (%)	Weight (g)	Moisture (%)	Weight (g)	Moisture (%)	Weight (g)	Moisture (%)
20	108	24	110	26	110	27	113	29
30	110	22	113	25	111	26	114	28
40	111	21	115	23	112	25	118	26
50	113	21	116	22	116	23	122	24
60	114	20	118	20.5	117	22	123	22
70	115	19.5	118	20	120	20	124	21
80	116	19	119	19.2	121	19.75	125	20.5
90	114	18	117	19	119	19.5	123	20
100	113	18	116	19	117	19.2	121	19.5

3.3 Effect of Baking Powder Addition on Baking Quality

3.3.1 Volume and Specific Volume

The increasing levels (1-8%) of baking powder affected the cakes volume. In general, it is observed that the volume of the cake increased vigorously up to 4% of addition and after that the volume of the cakes increased at a slow rate (Table 5). It could be seen that the highest volume resulted at 7% addition of baking powder and afterwards decreased progressively in all types of cake. The cake S_3 showed the highest cake volume among the samples at all levels of baking powder addition. At low levels of baking powder addition (1% or 2%) the cake of all samples showed not only low volume but the cakes were also under aerated, heavy and shrank on cooling. Although all the samples provided highest volume at 7% level of baking powder addition but also caused the cake to expand beyond the holding powers of the structural ingredients (such as flour and egg) and resulted in collapse structure giving a flat topped cake with an open texture. The cakes obtained at this level of baking powder addition (7%) resulted in adverse effects on the quality of the cake such as weakens the cake structure by breaking the gluten network, creating larger cell and by resulting an acid taste. The acceptable quality of cakes was obtained at 4 to 6% levels of baking powder addition in terms of texture, cell structure, air holes, flavor, shape and size of the crumb grain and color of both crumb and crust of the cake (Table 5). In case of specific volume, all the cake samples with different levels of baking powder addition from 1% to 8%, the maximum specific volume (2.14cc/g) of the cake sample S_3 attained at 7% baking powder addition, which was the highest among the samples. On the other hand, least specific volume (1.06cc/g) of the cake sample S_4 attained at 1% addition of baking powder that was lower than of all other samples with control sample. It was found that sample S_1 attained higher specific volume than all other samples except S_3 at 7% baking powder addition.

Table 5. Effect of various level of baking powder on volume and specific volume of plain cakes.

Baking powder level (%)	Cakes samples							
	S_1		S_2		S_3		S_4	
	Volume (cc)	Specific volume (cc/g)	Volume (cc)	Specific volume (cc/g)	Volume (cc)	Specific volume (cc/g)	Volume (cc)	Specific volume (cc/g)
1.0	162	1.27	165	1.25	170	1.26	155	1.06
2.0	185	1.55	190	1.52	192	1.50	180	1.36
3.0	210	1.81	220	1.85	225	1.85	200	1.6
4.0	214	1.86	226	1.91	230	1.90	208	1.70
5.0	220	1.94	231	1.97	238	2.01	215	1.77
6.0	225	1.99	235	2.00	243	2.05	221	1.85
7.0	230	2.05	240	2.08	249	2.14	228	1.93
8.0	227	2.03	236	2.07	244	2.12	225	1.92

Table 6. Effect of various level of baking powder on weight and moisture content of plain cakes.

Baking powder level (%)	Cake samples							
	S_1		S_2		S_3		S_4	
	Weight (g)	Moisture Content (%)	Weight (g)	Moisture Content (%)	Weight (g)	Moisture Content (%)	Weight (g)	Moisture Content (%)
1.0	128	22	132	23	135	23	140	24
2.0	120	20	125	21	128	22	132	23
3.0	116	19	119	19.2	122	19.75	125	20.5
4.0	115	19	118	19	121	19.5	123	19
5.0	113	18	117	18	118	19	121	18
6.0	113	17	117	17	118	18	119	18
7.0	112	18	115	18	116	17	118	17
8.0	111	16	113	17	115	18	117	17

3.3.2 Weight and Moisture Content

The effect of various levels of baking powder on the cake weight and moisture content are illustrated in Table 6. The addition of baking powder decreased the cake weight and moisture content. For sample S_1 the weight of the cakes varied significantly from 111 to 128 g being highest for 1% levels of baking powder addition and the lowest for 7% baking powder addition. The maximum moisture content (22.0%) of sample S_1 yielded at 1% baking powder addition, while the minimum moisture content (16.0%) was found at 8% baking powder addition, which was lower than of other samples. This may due to the gas produced in the dough during proofing by the amount of baking powder.

The moisture content of the cake containing coconut flour ranged from 17 to 24% and cakes weight varied from 113 to 140 g. The highest weight and moisture content was observed in cake sample S_4 at 1% baking powder addition, while the lowest weight and moisture content was observed in cake sample S_2 at 8% baking powder among the coconut cakes. At low levels of baking powder addition, all cake samples provided low volume, under aerated, heavy and shrank on cooling. Thus the moisture content and the weight increased in cake. Higher level of baking powder addition caused the cake to expand and resulted to decrease the moisture content and the weight of cake.

3.4 Sensory Evaluation of Plain Cake

The mean score for color, flavor, texture and overall acceptability preference are presented in Table 7. It revealed that there were significant ($P<0.05$) differences in flavor, texture and color acceptability among the cakes. The finding indicates that the color of S_3 cake had the maximum score (8.0), which indicates that the cake is more acceptable in color preference than the others. Similarly, the maximum score (7.9) was observed in S_3 cake in terms of texture preference. In contrast, flavor attribute showed that although S_4 cake received the maximum score (8.5), cake $S_3$was very close to S_4 cake, which received the score about 8.4. Singh et al. (2012) reported that flour replaced (20%) by corn bran resulted in cakes with acceptable sensory scores based on texture, taste and overall acceptability of the cakes. However, the overall acceptability preference was performed and the results revealed that the cake S_3 and S_2 were equally acceptable, but cake S_3 secured the highest score (8.1) to overall acceptability among the cakes (Table 7).

Table 7. Mean sensory scores of plain cakes.

Cake type	Sensory attributes			
	Color	Flavor	Texture	Overall acceptability
S_1	7.5^{AB}	7.6^{B}	7.7^{A}	7.4^{BC}
S_2	7.8^{A}	7.9^{B}	7.8^{A}	7.7^{AB}
S_3	8.0^{A}	8.4^{A}	7.9^{A}	8.1^{A}
S_4	6.8^{B}	8.5^{A}	6.8^{B}	7.1^{C}
LSD ($P<0.05$)	0.7381	0.4864	0.4933	0.5521

4. CONCLUSION

This study confirmed that incorporation of different levels of coconut flour to the cake formulation has improved the cake quality attribute nutritionally protein, minerals, fat and crude fiber and physically flavor, color, texture, volume, specific volume and weight. The cake type S_3 (20% coconut flour) is found more acceptable in terms of physicochemical properties of cake. Sensory evaluation proved that S_3 secured the highest score in color, texture and overall acceptability. The addition of sugar (20-100%) increased the cake weight and decreased the moisture content of cake, in which the cake volume and specific volume increased up to 80% sugar addition in cake dough. In contrast, the increase of baking powder decreased the cake weight and moisture content, but increased the cake volume and specific volume up to 7% baking powder addition. The 80% sugar and 7% baking powder addition in the dough showed the highest cake volume, specific volume and weight and low moisture content in all types of cake. However, S_3 cake incorporated with 20% coconut flour showed the maximum outputs and better acceptability, when the sugar and baking powder addition were around 80% and 7%, respectively, in the dough. In conclusion, good quality coconut flour cake could be processed for improved nutritional value and other aspects.

ACKNOWLEDGEMENT

The authors would like to thank to the Department of Food Technology and Rural Industries, Bangladesh Agricultural University, Mymensingh-2202, Bangladesh for supporting this research project.

REFERENCES

1. T. A. Nasrin, 2002, Master Thesis: "Development of coconut products using fresh and dried coconut kernel". Department of Food Technology and Rural Industries, Bangladesh Agricultural University, Mymensingh-2202, Bangladesh.

2. P. T. Trinidad, H. V. Divinagracia, C. M. Aida, C. A. Faaridah, S. M. Angelica, T. C. Modesto, C. A. Askali, A. S. Loyola, and D. B. Masa, 2001, "Coconut flour from residue: A good source of dietary fibre". Indian Coconut J. 32(6), 9-13.

3. R. N. Arancon, 1999, "Coconut flour". Coco Info International, 6(1), 1-8.

4. N. Hesso, C. Garnier, C. Loisel, S. Chevallier, B. Bouchet, A. Le-Bail, 2015a, "Formulation effect study on batter and cake microstructure: Correlation with rheology and texture". Food Struc. 5: 31-41.

5. N. Hesso, C. Loisel, S. Chevallier, A. Le-Bail, Queveau, B. Pontoire, P. Le-Bail, 2015b, "Monitoring cake baking by studying different ingredient interactions: From a model system to a real system". Food Hydro. 51: 7-15.

6. Sindhuja, M. L. Sudha, and A. Rahim, 2005, "Effect of Incorporation of Amaranth Flour on the Quality of Cookies," European Food Res. Technol. 221 (5), 597-601.

7. M. L. Sudha, R. Vetrimani, and K. Leelavathi, 2007, "Influence of Fibre from Different Cereals on the Rheological Characteristics of Wheat Flour Dough and on Biscuit Quality," Food Chem. 100 (4), 1365-1370.

8. M. N. Khan, R. D. Hagenmaier, L. W. Rooney and K. F. Mattil, 1976, "High Fibre Coconut Products for Baking Systems". Baker's Digest. 50, 19-25.

9. P. Corpuz, 2004, "Philippine Oil Seeds and Products Annual". Global Agricultural Information Network, USDA Foreign Agricultural Service, http://www.cocoscience.Com/pdf/descriptive_sensory_evaluation.pd.

10. F. A. S. Dairo, 2007, "Evaluation of Protein Replacement Value of Sun Dried and Oven Dried Coconut Oil Meal and Fermented Coconut Oil Meal in Rats". Inter. Journal Agr. Res. 2, 246-253.

11. J. H. C. Atkins, 1971, "Mixing requirement of baked products" Food. Manuf. 47: 120.

12. C. L. Rajchel, M. E. Zabik, and E. Everson, 1975, "Wheat bran and middlings: A source of dietary fiber in banana, chocolate, nut and spices cakes". Bakers Dig., 49, 27-30.

13. AOAC, 2005, "Official Methods of Analysis, 12th edition", Association of Official Analytical Chemists, Washington, D. C. USA.

14. D. Pearson, 1970, "The Chemical Analysis of Foods, 7th edition". Churchill Lavingstone, New York.

15. D. B. Ott, 1987, "Applied Food Science Laboratory Manual". Michigan State University. Pergramon press. USA.

16. W. T. W. Leung, M. S. Ritva Rauanheimo Butrum, and B. S. Fiora Huang Chang, 1972, Food and Agriculture Organization of the United Nations, Food Policy and Nutrition Division.

17. Y. Bae, H. I. Lee, A. Ko and H. G. Lee, 2013, "Substituting Whole Grain Flour for Wheat Flour: Impact on Cake Quality and Glycemic Index". Food Sci. Biotechnol. 22(5): 1301-1307. DOI 10.1007/s10068-013-0216-4.

18. M. Singh, S. X. Liu, S. F. Vaughn, 2012, "Effect of corn bran as dietary fiber addition on baking and sensory quality". Biocat. Agr. Biotechnol. 1:348–352.

19. Y. Pomeranz, 1980a, "What? How Much? What function? In bread making", Cereal Foods World, 25: 656-662.

20. M. M. Karaoglu, H. G. Kotancilar, 2009, "Quality and textural behaviour of par-baked and rebaked cake during prolonged storage". Int. J. Food Sci. Technol. 44: 93-99.

21. Kim, S. M. Lee, I. Y. Bae, H. G. Park, H. G. Lee and S. Lee, 2011. "(1→3)(1→6)-β- Glucan-enriched materials from Lentinus edodes mushroom as a high-fibre and low-calorie flour substitute for baked foods". J. Sci. Food Agr. 91: 1915-1919.

22. B. K. Rahut, S. Hossain, M. H. R. Bhuiyan and M. Shams-Ud-Din, 2012, "Study on composite flour cake and palmyra palm incorporated cake". Bangladesh Res. Pub. J. 7(4), 378-385.

23. S. P. Cauvain and J. A. Cyster, 1996. "Sponge Cake Technology", CCFRA Review No. 2. CCFRA, Chipping, Campden, UK.

4

EFFECT OF HIGHLY AERATED FOOD ON EXPECTED SATIETY

Juan-Carlos Arboleya[a,], Maruxa García-Quiroga[a], Daniel Lasa[b], Oswaldo Oliva[b], Andoni Luis-Aduriz[b]

[a] AZTI-Tecnalia Food Research Institute, Parque Tecnológico de Bizkaia, Astondo Bidea, Edificio 609, 48160 Derio, Bizkaia, Spain

[b] Mugaritz Restaurant, Otzazulueta baserria, Aldura aldea 20 zk. – Errenteria, 20100 Gipuzkoa, Spain

ABSTRACT

This work shows a practical way to design satiating new products by a real interaction of science and cooking. From the initial idea of the chef, a highly aerated product was designed to prove that the feeling of fullness starts before food is eaten, at the point when the food is just being viewed by the consumer. Mixtures of food-grade silica particles, methylcellulose (MC) and ovalbumin (OA) were used to get better distribution of air and to increase volume. Silica particles at a concentration of 0.3 wt% , mixed with MC (0.5 wt%) and OA (1 wt%) showed higher surface activity and viscoelasticity at the surface than the isolated ingredients. This mixture also showed the highest foam capacity and foam stability compared to the mixtures with none or 0.4 wt% of silica. Highly aerated structures were made by using the mentioned results. To verify the idea of having higher expected satiety with a highly aerated product, consumer study was performed. Subjects reduced their intake when a more-aerated sample was served compared to a less aerated sample.

KEYWORDS

Proteins; Methylcellulose; Silica particles; Foaming properties; Satiety

1. INTRODUCTION

Food industry must have a strong focus on delivering innovation to meet market and consumer trends in health, texture, nutrition and targeted delivery solutions (Berry, 2008). Food developers must remember the following consumer demands: Is it good for me? Is it good for the world? Does it taste good? Does it consider gastronomic culture? Does it give me an extra-value? Chefs could play a key role by inspiring food industry with new developments and by manufacturing healthy and tasty food products. In fact, the creation of healthier, tastier and innovative food is one of the most important concerns of chefs nowadays. From a nutritional point of view, for example, science-based cooking can contribute to provide certain nutrients and other food components, which could confer healthy aspects to the dishes and menus (Navarro et al., 2012).

Common strategies for promoting healthy eating habits are controlling portion size and reducing the energy density of the meal (Hazen, 2007). The mechanisms controlling appetite and hunger are complicated but the study of these mechanisms could help get scientific evidence that could assist consumers in making healthier food choices. There are two different processes related to food intake. Firstly, satiation, that refers to the act of stopping food intake, an immediate reaction to the ingestion of food. During satiation, it seems clear that sensory (taste, volume, texture...) and cognitive factors will play an important role (De Graaf, 2012). Secondly, satiety, that refers to motivations to eat in between meals (Blundell et al., 2010), is body's response to the availability of nutrients from food that have been already digested and processed. From those close but different concepts, it has been reported that deficiencies in satiation seem to show more connections to obesity and binge eating than actual satiety (Kissileff, 1995 and Spiegel et al., 1989). Therefore, enhancing satiation could restrict the daily food intake and the desire of overeating, contributing to body weight control

(Hoad et al., 2004). Considering that satiation is partially governed by sensory factors (Blundell et al., 2010), it would be interesting to develop a product focused on the satiating effect of sensory perception, such as the visual factor. Consumer's sight influences the quantity of consumption, leading them to be less influenced by physiological signals of satiation. Wansink (2005)showed that manipulating visual signals of how much is eaten influences further intake:*people use their eyes to count calories and not their stomachs.*

There are many methods for the development of products related to this effect (satiation), which are based on lowering caloric density or modifying structure. For instance: the use of non-caloric ingredients, immobilizing high quantities of water and incorporating air as small dispersed bubbles (Zúñiga and Aguilera, 2008). Recent published research has highlighted the link between applied material characteristics like viscosity or gel performance of polysaccharides in food matrices and the effect of gastric emptying and satiety (Knarr et al., 2012). Methylcellulose (MC), for example, could be a good alternative, because it is a modified cellulose fiber that produces viscous solutions in the gastrointestinal tract. Furthermore, at the 244th National Meeting & Exposition of the American Chemical Society, a new product was presented based on modified methylcellulose that could present satiety properties (Huettermann, 2012).

Fixing air in food could be another alternative to design satiating products. In recent years, new aerated foods from the market are perceived as lighter in terms of calories, thus satisfying consumer needs. However, introducing a gas phase into a food matrix not only affects its texture and firmness making the product lighter, but also changes the appearance, color and mouth-feel (Campbell and Mougeot, 1999). So, it is essential to revise this aspect because these products are not only consumed for health purposes, but also for enjoyment. Aerated foods may provide a sense of fullness in comparison to the non-aerated food. There is some evidence that food intake is influenced by both the weight and volume of foods. It has been reported that increasing the air content may be an effective strategy to reduce energy intake from energy-dense products, like snacks (Osterholt et al., 2007). In this study, consumers ate 21% less weight and energy of the more-aerated snack than the less-aerated snack. This work presented the product with the same volume for both cases, so it is impossible to know whether consumers expected to be satiated before eating the product, just by sight signals. To the best of our knowledge, there is no literature at all describing the relation between expected satiety (from initially visual observation) and product volume.

The main objective of the present study was to design a significantly high-aerated product. This scientific study, together with chefs' gastronomic knowledge, would be used to fix the initially aerated structure, to get better distribution of air and increase volume, in order to check the perception of being full before eating. It was then hypothesized that the most-aerated product might give higher expected satiety than the less-aerated product.

2. MATERIALS AND METHODS

2.1 Materials

Methyl cellulose (Methocel Premium A15, mean molecular weight 14 kDa, methyl substitution between 27.5% and 31.5%) was obtained from The Dow Chemical Company (Midland, TX). Albumin from chicken egg white (A-5504) was purchased from Sigma-Aldrich (98%) and fumed synthetic amorphous silica AEROSIL® was purchased from Evonik Industries.

Two methodologies were used to get samples with different aeration. Two recipes were whipped at different speed to get structures of diverse volumes. The first one, the high aerated product (HA) was formed by pasteurized egg white, methyl cellulose (MC), amorphous silica and maltodextrin, which helped in the reinforcement of the foam structure. This dough was whipped in a Kitchen Aid Fagor® for 5 min at maximum speed. The second sample, the low aerated product (LA) was made just by pasteurized egg white, methyl cellulose and maltodextrin, and was whipped at the 4th speed for 20 min. After whipping, visually different samples were obtained with both methodologies. The final presentation of the dish ends with 15 g of each type of foam put in a Pirex® bowl.

2.2 Surface tension

Surface tensions at the air–water interface of protein solutions were measured by using an FTA200 pulsating drop tensiometer (First Ten Ångstroms, USA). The capillary drop was formed within an environmental chamber at room temperature, in which standing water increased the relative humidity to minimize drying effects. All measurements were made at room temperature (≈20 °C). Surface tension was monitored at room temperature for 30 min.

2.3 Surface rheology

Surface shear rheological measurements were carried out to study the mechanical and flow properties of adsorbed layers at fluid interfaces, which were sensitive to surface structure and composition (Ridout et al., 2004). Experiments at the air–water interface were made using a stress controlled rheometer, AR2000 Advanced Rheometer (TA Instruments) and an aluminum bicone (diameter 60 mm, angle cone 4:59:13) as measuring geometry. The surface rheological response in 50 mL solution was tested by oscillation mode within the range of linear viscoelastic region at a frequency and strain of 0.1 Hz and 0.014, respectively. Measurements were performed at room temperature for 30 min.

2.4 Foaming properties

Foam production was achieved by using a Foamscan TM apparatus (Teclis-ITConcept, Longessaigne, France), whose principle is to foam a 10 mL solution by gas sparging (N_2) through a porous glass frit (flow of gas: 45 mL min^{-1}; porosity 16–40 μm). The amount of liquid incorporated in the foam and the foam homogeneity are followed by measuring the conductance in the cuvette containing the liquid and at different heights in the column by means of electrodes. Bubbling was stopped after 80 s. Drainage of the foam was followed via conductivity measurements at different heights of the foam column. A pair of electrodes at the bottom of the column was used for measuring the quantity of liquid that was not in the foam, while the volume of liquid in the foam was measured by conductimetry. After turning off the gas, the foam height was read from the calibrated markings on the column. The time for the collapse of the foam to half its initial value was measured to express the foam stability. Two parameters were determined as a measure of foaming capacity. The overall foaming capacity (FC) was determined by measuring the maximum volume reached by the foam at the end of sparging. Foam stability was characterized by half-life of foam $t_{1/2}$ (s), the time for draining half of the initial volume of the foam (Eq. (1))

$$t_{1/2} = (k_2 V_o)^{-1} \quad (1)$$

3. CONSUMER STUDY

3.1 Subjects

There were **25 subjects** (M/F: 6/9, age: 35±6.0 and BMI: 22.9±2.7 kg/m^2) who were students or employees at the Azti-Tecnalia Food Research Institute. Participants completed an inclusion questionnaire, in which body weight and height were reported. Exclusion criteria were: lack of appetite, following an energy-restricted diet or change in body weight >5 kg during the last 2 months, stomach or bowel diseases, diabetes, thyroid disease or any other endocrine disorder and hypersensitivity for food products under study. Subjects were unaware of the aim of the research.

3.2 Procedure and experimental design

Regarding the consumer behavior, subjects were instructed to consume the same breakfast and not eat anything else and only drink water or weak tea, 2 h before lunch started. Moreover, they were asked to refrain from drinking 1 h before the test started. To make sure that the subjects would eat till they felt satiated; they were instructed not to eat 1 h after the test. Each subject was seated on a separate table with a sample bowl. During break (10.30 h–11.00 h), subjects had free access to water. Subjects had to indicate the hour when they started and when they finished consumption of the samples, so that the eating time was recorded. They were instructed to terminate consumption when they felt they had enough. The mean initial temperature of the aerated product was 20 °C.

We manipulated two stages: (1) expected satiation and (2) the acceptability and the oral exposure time at the end of the oral eating process.

In the first stage, participants rated the expected satiation on a 100 mm line scale anchored from "not full at all" to "extremely full" to the question "how filling would you expect this portion of food to be?". This procedure was repeated for the different samples and presentation order was randomized across participants. Average expected satiation values were calculated for all structure items by averaging across participants.

In the second stage, oral perception at the end of the oral eating process, subjects were served a sample of 15 g to rate after tasting the sample (Fig. 1). When they finished the product, it was rated for overall acceptability according to the scale VAS.

For each individual, two samples were analyzed in two different sessions with a week's gap between them. Twelve consumers were invited to evaluate the LA sample and the rest of the group evaluated the HA sample in the first sessions. The same occurred in the second session, but in reverse.

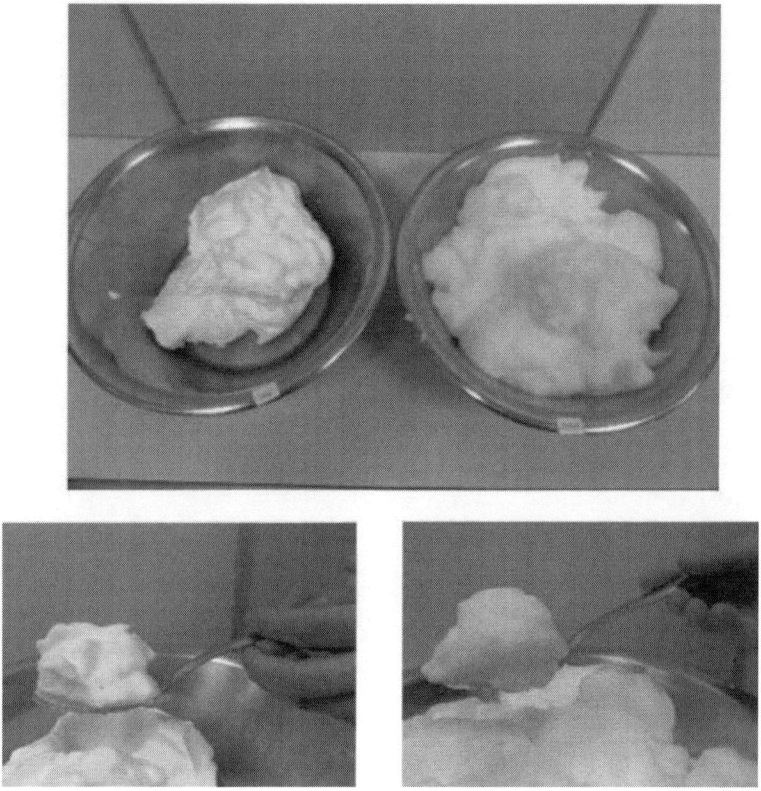

Figure. 1. High-aerated and low-aerated samples for sensory studies. Subjects were served a sample of 15 g.

3.3 Data analysis

Statistical analysis was carried out with the R 2.14.0 Programme (R Development Core Team), with the packages Commander, Agricolae, Sensominer and Factominer. One way Analysis of Variance (ANOVA) was employed to check for individual differences between two samples.

4. RESULTS AND DISCUSSION

4.1 Designing an aerated food product for the restaurant

Our starting point was that an aerated product could potentially achieve a reduction in caloric density and it could induce satiety by making novel gastronomic structures.

Another important issue for the chef was to create, literally, the sensation of lightness in a dish.

The creation of foams essentially requires the formation of fine bubbles. The main factor controlling droplet size is the interfacial tension (Wilde, 2000). Apart from the considerable interest in the effects of silica on human health (Martin, 2007), these partially hydrophobic particles are also known to stabilize the interface even in the absence of any added surfactant (Aveyard et al., 2003), and extend storage stability (Castro et al., 2006). The development of surface tension in time, upon creation at the air–water interface of two different samples, was studied by using the pendant drop technique using an essential protein solution (OA) and a surface active polysaccharide (MC). The surface tension of MC and OA solutions are shown in Fig. 2a and b, respectively. Isolated MC samples lowered the surface tension more rapidly than isolated OA even with half the concentration. Adding 0.3 wt% of silica particles in a constant concentration of MC (0.5 wt%) and OA (1 wt%) seemed to lower the surface tension more rapidly than isolated ingredients without silica particles whereas at 0.4% of silica the system was less surface-active (Fig. 2).

The surface rheology of adsorbed amphiphilic compounds or even particles has long been known to be important for the stabilization of foams. The impact of silica particles on the surface rheological properties of OA and MC solutions was then investigated. Fig. 3shows the shear elastic modulus of 0.5 wt% and 1 wt% solutions of MC and OA measured over a period of 30 min. It can be observed that G' at the presence of 0.3% silica particles increased more rapidly than isolated ingredients at short times followed by slow increase at longer times. The increase in surface rheology with time can be considered to be a result of silica particles adsorption at the interface, forming a much stronger interface. This synergy might be attributed to the ability of MC and OA to increase the silica particle contact angle on adsorption (Hunter et al., 2008). However, MC and OA with 0.4 wt% of silica showed less elastic interfaces than isolated ingredients and mixtures with 0.3% of silica. This increased silica concentration may lead to site competition between the particles and surface active molecules. Foaming properties (Fig. 4) seemed strongly correlated to the rate of adsorption at the air–water interface (Fig. 2) and thereby to the decrease of surface tension, giving a higher foaming capacity up to 0.3% of silica particles for both mixtures of MC and OA. The higher viscoelastic properties (Fig. 3), once again for both mixtures of MC and OA and up to 0.3% of silica particles, resist the shear stress resulting from liquid drainage and improve the foamability and foam stability under these conditions. A higher viscosity in the aqueous solution could retard the drainage in the foam. This does not seem to be the case because of the small differences in viscosity of the solutions made by isolated OA and MC compared to silica–OA–MC solutions (data not shown). Silica nanoparticles seem to provide stability in foams, especially related to disproportionation and bubble shrinkage thus giving much better results than with isolated proteins (Dickinson, 2010).

By using the knowledge acquired from the studies described above, the recipe was formed by ovoalbumin, methylcellulose and amorphous silica which helped in the reinforcement of the foam structure. Additionally, sugar was replaced by a syrup made by maltodextrin and water, which was inserted into the whipped mixture. Finally, the aerated structure was left in an oven at 55 °C to obtain the final structure. The final presentation of the dish includes different varieties of berries and only 2 g of the aerated structure (Fig. 5).

4.2 Checking the relation between aeration degree and expected feeling of fullness

Since the aim of the work is to assess the influence of aeration type on the intake of different samples, one important aspect was to know the influence of the intake, the oral exposure time and the acceptability about high and low aeration.

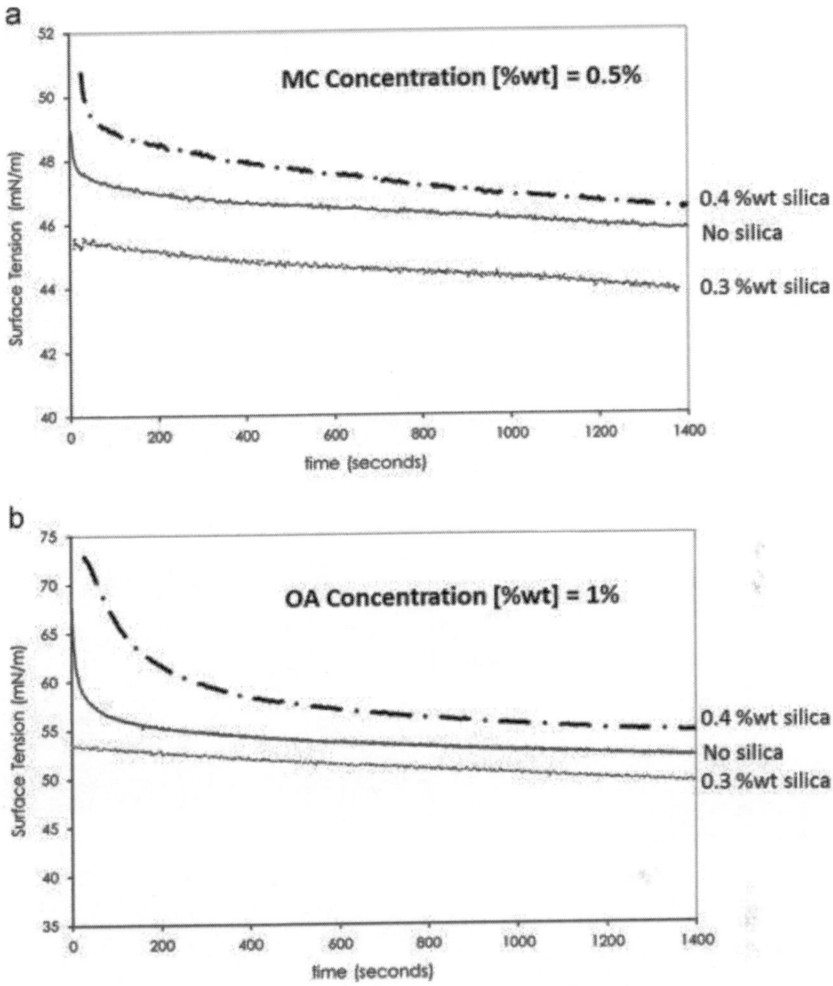

Figure. 2. Effect of silica concentration on the surface tension (taken at 30 min) versus time as measured by the pendant drop technique for (a) constant concentration of methylcellulose (MC) (0.5 wt%) and (b) constant concentration of ovalbumin (OA) (1 wt%).

Aeration degree of the product had a significant effect on the intake (p=0.03; Table 1). On the other hand, there was no significant effect either on the oral exposure time (p=0.63;Table 1) or on the acceptability (p=0.24; Table 1). However, the results should still be taken with care , since within the absolute error margin, the intake falls in the same window.

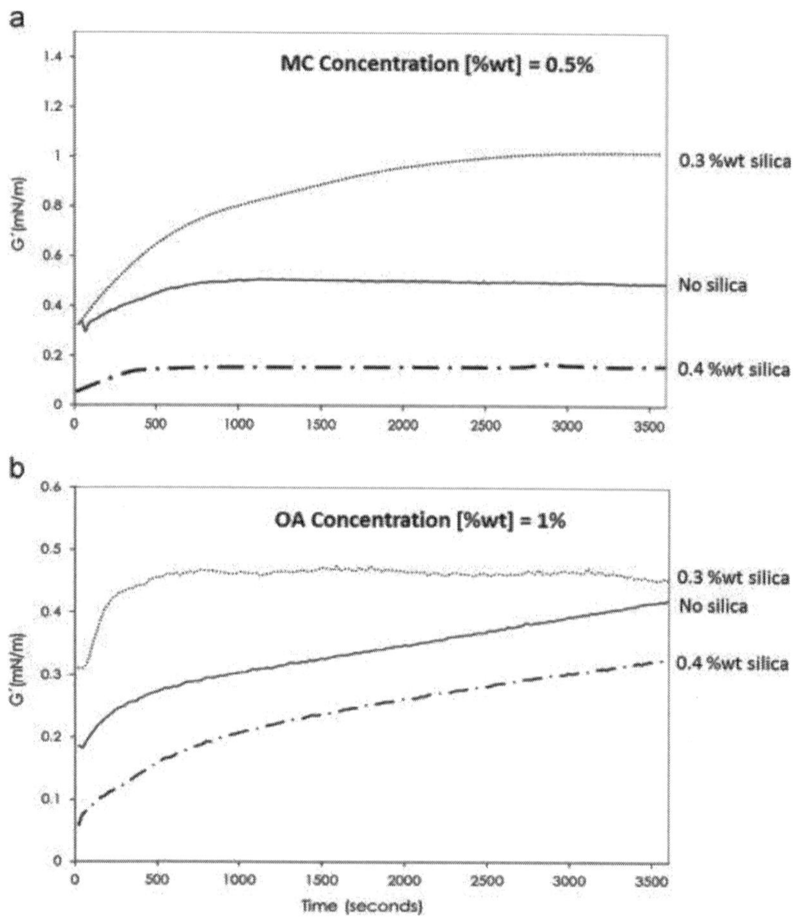

Figure. 3. Effect of silica concentration on the surface shear elastic modulus versus time for (a) constant concentration of methylcellulose (MC) (0.5 wt%) and (b) constant concentration of ovalbumin (OA) (1 wt%).

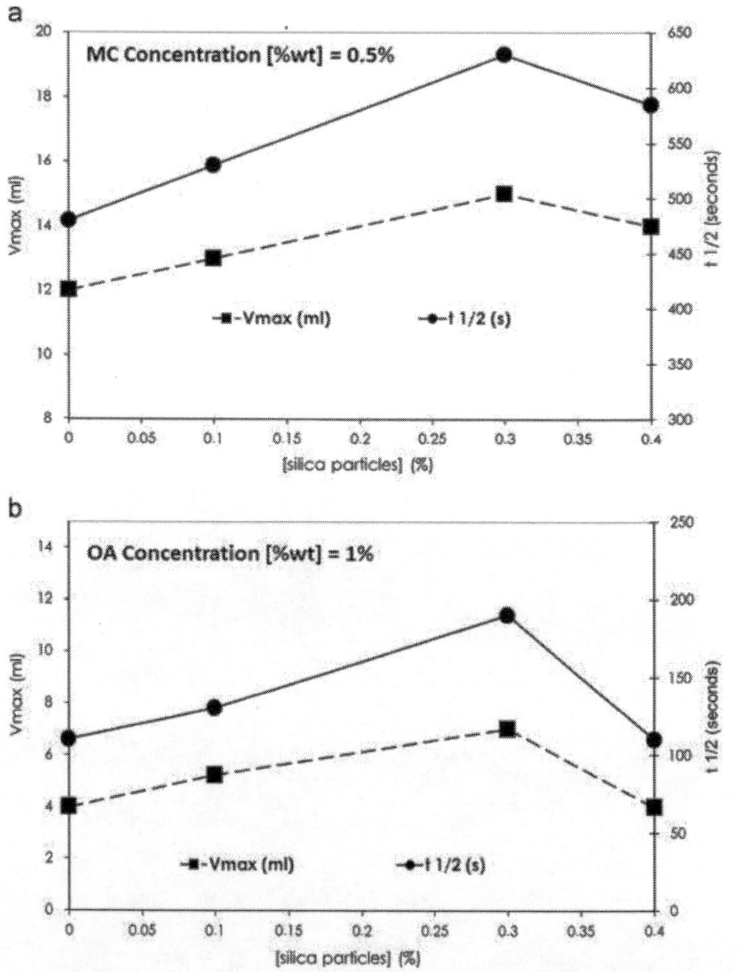

Figure. 4. Effect of silica concentration on foaming capacity and foaming stability for (a) constant concentration of methylcellulose (MC) (0.5 wt%) and (b) constant concentration of ovalbumin (OA) (1 wt%).

According to Hogenkamp et al. (2011), the effect of satiety expectations on actual intake may be important to better understand our regulation of food intake. The ratio of expected satiety of high and low aeration was 56.12 mm and 44.52 mm (scale from 0 at 100 mm) respectively. This means that the aeration degree of the product had a significant effect on the expectation (p=0.03; Table 1). The study clearly shows that a difference in the air content of a food affected the volume consumed when two types of samples were served and consumed *ad libitum* intake with same weight and calories.

Figure. 5. An aerated product that induces a sense of fullness based on the principle that satiation begins when you first look at the product: *Evoking a spring morning*. Photos by Jose Luis Lopez de Zubiria-Mugaritz.

5. CONCLUSIONS

Silica particles at a specific concentration and mixed with ovalbumin and methylcellulose showed higher surface activity and more viscoelasticity at the surface than the isolated ingredients. The highest foam capacity and foam stability were obtained by adding 0.3 wt% of silica particles in a constant concentration of MC (0.5 wt%) and OA (1 wt%) which is related to the studied interfacial properties.

Table 1. Results of the means, standard deviation and one way ANOVA (aeration) of the intake, oral exposure time, acceptability and expected satiety.

Variables	HA	LA	*F*	*p*-Value
Intake (g)	6.74 $\pm$ 4.50	9.32 $\pm$ 3.59	5.03	0.03
Time intake (min)	2.50 $\pm$ 1.08	2.60 $\pm$ 1.28	0.22	0.63
Acceptability (mm)	48.40 $\pm$ 20.14	53.60 $\pm$ 18.00	0.92	0.34
Expected satiety (mm)	56.12 $\pm$ 17.74	44.52 $\pm$ 19.23	4.91	0.03

This study permitted to produce highly aerated structures to arrive at the design of new dishes mainly focused on producing high levels of expected satiety.

Subjects reduced their intake when a more-aerated sample was served. These results confirm and extend the findings from previous studies indicating that the amount of air incorporated into food can affect the energy intake (Rolls et al., 2000 and Osterholt et al., 2007). These results open an interesting and promising path for the design of food for wellness. Further research on satiety and satiation should include chefs for a better understanding of these complex mechanisms.

REFERENCES

1. Aveyard, R., Binks, B. P., Clint, J. H., 2003. Emulsions stabilized solely by colloidal particles. Adv. Colloid Interface Sci.100,503–546.

2. Berry, D., 2008. Texture without the calories. ⟨http://www.foodproductdesign. com⟩ (5-5-2008). Internet Communication.

3. Blundell, J.,de Graaf, C., Hulshof, T., Jebb, S., Living stone, B., Lluch, A., Mela, D., Salah,S.,Schuring,E.,VanDerKnaap,H.,Westerterp,M., 2010. Appetite control:methodological aspects of the evaluation of foods. Obes. Rev.11(3),251–270.

4. Campbell, G.M., Mougeot, E.,1999. Creation and characterization of aerated food products.Trends FoodSci. Technol.10(9),283–296.

5. Castro, I.A., Motizuki, M.,Murai, H.,Chiu, M.C.,Silva, R.S.S.F., 2006. Effect of anti caking agent addition and headspace reduction in the powdered- drink mix sensory stability. J.Food Qual.29(3),203–215.

6. De Graaf, C.,2012. Texture and satiation: the role of oro-sensory exposure time. Physiol.Behav.107(4),496–501.

7. Dickinson,E.,2010.Food emulsions and foams: stabilization by particles. Curr. Opin. Colloid Interface Sci.15(1–2), 40–49.

8. Hazen, C,2007.Servingupcalorie-controlledentrées. ⟨http://www. foodpro ductdesign.com/articles/463/7ah3110372951050.html⟩ (31-10-2007). Internet Communication

9. Hoad,C.L.,Rayment,P.,Spiller,R.C.,Marciani,L.,Alonso,B.C.,Traynor,C., etal.,2004.Invivoimagingofintragastricgelationanditseffectonsatietyi n humans.J.Nutr.134,2293–2300.

10. Hogenkamp,P.S.,Stafleu, A.,Mars,M.,Brunstrom,J.M.,Graaf,C,2011. Texture, not flavor, determinesexpectedsatiationofdairyproducts. Appetite 57,635–641.

11. Huettermann,C.,2012.Newformoflong-usedfoodingredientfor "anti-hunger" yogurts, smoothies.ACSPressRelease. ⟨http://portal. Acs org/ portal/acs/corg/content?_ nfpb=true&_ pageLabel= PP_ ARTICLEMAIN& node_id=222&content_id=CNBP_030562&use_ sec=true&sec_url_var= region1&__uuid=8e0e953b-ee68-4b65-a034-929465d6f401⟩.

12. Hunter, T.N., Pugh,R.J.,Franks,G.V.,Jameson,G.J,2008.Theroleof particles instabilising foam sand emulsions. Adv.ColloidInterfaceSci. 137, 57–81.

13. Kissileff, H.R.,1995. Inhibitions of eating in humans: assessment, mechanisms, and disturbances.Appetite,225–299.

14. Knarr, M., Adden, R.,Anderson,W.H.K.,Hübner-Keese,B.,2012.Characterization of in-vitrogelperformanceofnovelMCwithrespecttothe suitability forsatietyapplications.FoodHydrocoll.29,317–325.

15. Martin, K.R., 2007. The chemistry of silica and its potential health benefits. J. Nutr. Health Aging11(2),94–98.

16. Navarro, V.,Serrano,G.,Lasa,D.,Luis-Aduriz,A.,Ayo,J.,2012. Cooking and nutritional science: gastronomy goes further. Int. J. Gastron. FoodSci. 1 (1),37–45.

17. Osterholt, K.M.,Roe, L.S.,Rolls,B.J,2007.Incorporationofairintoasnack food reducesenergyintake.Appetite48(3),351–358.

18. Ridout, M.J., Mackie, A.R.,Wilde,P.J.,2004.Rheologyofmixed β-casein/β- lactoglobulin films attheair–water interface. J.Agric. Food Chem. 52(12), 3930–3937.

19. Rolls, B.J., Bell, E.A., Waugh, B.A., 2000. Increasing the volume of a food by incorporating airaffectssatietyinmen.Am.J.Clin.Nutr.72, 361–368.

20. Spiegel, T.A.,Shrager,E.E.,Stellar,E.,1989.Responsesofleanandbese subjects topreloads,deprivation,andpalatability.Appetite13,45–69.

21. Wansink, B. ,2005. Bottom lessbowls: why visual cues of portions izemay influence intake. Obes.Res.13(1),93–100.

22. Wilde, P.J., 2000.Interfaces: the irroleinfoamandemulsion behavior.Curr. Opin. ColloidInterfaceSci.5, 176–181.

23. Zúñiga, R.N. ,Aguilera,J.M.,2008.Aeratedfoodgels:fabricationand potentialapplications.TrendsFoodSci.Technol.19(4),176–187.

5

BOLLITO MISTO NON-BOLLITO

Enrico Vignoli, Yoji Tokuyoshi, Massimo Bottura

Osteria Francescana, Via Stella, 22, 41100 Modena, Italy

ABSTRACT

An improvement in nutritional and tenderness values, among others, was achieved in a traditional northern Italian recipe, the so-called *bollito misto* (a mixture of boiled meat) by integrating low temperature cooking. This technique can be used within the traditional recipe and therefore considerably enhance the sensory and the nutritional properties of the dish. This process would get a much tastier meat, preserving meat quality and keeping sauce flavors. This recipe update clearly gives a step forward from a traditional formula to an improved version. These "new traditions" should give new hints to the normal food that people usually consume.

KEYWORDS

Sous vide; Low temperature cooking; Boiled; Bollito

1. INTRODUCTION TO THE CULINARY CONCEPT

Tradition and innovation walk hand in hand. They need one another. When technical knowledge is needed to move forward, the access to the scientific knowledge is required. A way to develop new products from Osteria Francescana is always based on tradition. From this point on, a very traditional dish can be taken from the past and brought to the present to incorporate an innovation. This should be done keeping a strict connection to territory and tradition (Sultano and Bottura, 2006). Therefore, understanding and interiorizing tradition may open the possibility of modifying it, as it has always been modified in the history of men.

A very popular and traditional dish in the North of Italy is the so-called *bollito misto*(boiled) which is based on the use of boiled low quality meat (more difficult to soft by cooking) (Fig. 1). Lower class at the area of Modena used to boil pieces from the hardest parts of the veal – *vitello* – cheek, tongue, tail, belly and head, which contains high contents of fat and collagen (Boni, 1971). It also used the *cotechino* – pork – a kind of salumi (big sausage) made by wrapped pieces of cured meat and it is formed mainly by skin, head and cheek (Molinari Pradelli, 1999). Once meat is boiled, the broth is used for other applications, like the preparation of pasta. The boiled meat is the core of this recipe, although it showed very poor quality as it is overcooked, exhausted without any juice inside and completely dry and tasteless. For other areas, different parts are used (Carnacina and Veronelli, 1960). For instance, in Piemonte, *the gran bollito* is made by seven pieces of beef and veal (neck, belly, rump with thigh muscle, shin, shoulder, *fiocco di punta, cappello del prete*). In other areas, this dish contains other pieces of meat (tongue, head, cotechino, chicken, queue, trotter and hand roll). Traditional way to serve is accompanied with several kind of vegetables according to the season (usually potatoes, spinach, carrots, onions, cauliflower, green beans and roots) and seven kind of sauces: mustard (fruit, sugar and salt) in Cremona and Mantova; green sauce (made by garlic, anchovies, vinegar, cappers, oil, parsley and bread soaked in vinegar) in Emidia Romagna and Piemonte (Artusi, 1891), red sauce, honey sauce, horseradish sauce and mash in other areas. All of them hide any taste of meat.

Figure. 1. Ancestral Italian dish: *bollito misto.*

The enhancement of the quality in these poor parts of the veal and pork brings us to profoundly know this material by collaborating with the best meat producers and breeders. Thus, different animals from different producers are cautiously chosen in this regard. *Testina* for example, which is usually rolled and wrapped, is normally unpacked to take all the pieces one by one. This preliminary treatment of the meat is crucial to add value to this product. Therefore, using new culinary techniques into a very traditional dish can give a significant improvement of the original dish. *Sous vide* cooking is the process of cooking vacuum sealed food in a low temperature water bath to achieve a desired texture and it could be very helpful for the enhancement of this product (Roca and Brugués, 2003).

The aim of this culinary concept is to obtain a much tastier meat after using low temperature cooking (*sous vide cooking*) in sealed bags, preserving meat identity and keeping sauce flavors. This objective must be done without forgetting the traditional concept, the traditional ingredients and flavors and after all, the people. Sauces must also be kept in this new concept but it is needed to bear in mind that meat shows a more intense taste than the traditional recipe, so, these sauces work as a more subtle component. In summary, meat is tastier in this version and the sauces become a secondary elaboration.

2. CULINARY CONCEPT: DEFINITION

Tradition and innovation can coexist by using culinary techniques in popular and typical dishes, for instance, the so-called *bollito misto* from the North of Italy. Low temperature cooking can be integrated within the traditional recipe and therefore considerably enhance the sensory and the nutritional properties of the dish.

2.1 Description of the culinary process

2.1.1 Cooking meat

Sous vide cooking or low temperature cooking is used to achieve much more interesting flavor and texture of the treated meat. This technique relies on a precise, temperature controlled circulator. In this recipe, different types of meat are put into individual bags, each type of meat separately, with several types of vegetables, herbs (carrots, onions...) and poultry consommé. This would keep all the water soluble flavors in the meat apart from the vegetable and herbs flavors (in the traditional way, added herbs are poured into the water). The bags were vacuum sealed and cooked at a gentle temperature in a precisely controlled water bath (Fig. 2). This system will promote correct heat transfer and convection, two crucial parameters to reach the optimum point in meat texture and flavor. Sealed bags avoid meat to be contacted with water; thereby less dehydration will exist at the meat surface retaining flavors and developing tender textures (Baldwin, 2012). Osmosis process is considerably reduced keeping higher mineral content in the product. By using this method, it is possible to improve the tenderness of the meat comparing to the traditional way of cooking.

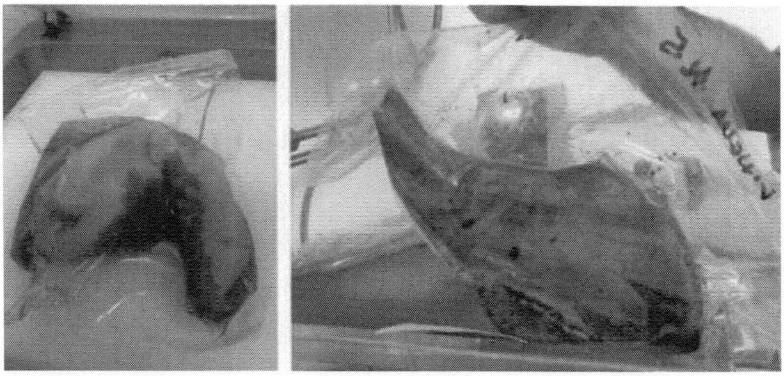

Figure. 2. Pieces of meat into vacuum sealed bags ready to cook in a controlled water bath.

Different temperatures and cooking times were selected to get a specific texture for each type of meat: cheek, tongue and belly (75 °C, 20 h), tail and *testina* (head) (75 °C, 36 °C),*cotechino* (pork) (63 °C, 15 h). Some of them, like the head or *testina* and tail, contain a higher content of collagen, therefore higher temperatures and longer cooking time must be applied to avoid a *gummy* sensation and get the best texture. On the other hand, the belly would need shorter cooking time. These parameters will also depend on the age of the animal.

Bearing in mind the traditional *bollito misto*, this culinary concept tried to maintain the characteristic flavor and taste that sauces grant to the dish but without hiding meat flavor. A new form of serving the sauce was proposed by making a foam, which contributes with a less intense but enhanced mouth feel. Foam was made by adding the broth from the sous vide bags, soy lecithin, salt and sugar to have a more direct flavor which remains the green sauce. Some ingredients from the green sauce are also in the base underneath, in a jelly base made with cappers and anchovies to strengthen the sensation of the green sauce. *Peperonata* (made with yellow and red pepperoni) was added to recover a sour sensation (Fig. 3). A first version of the improved traditional *bollito misto* was called *bollito misto non-bollito* (boiled non-boiled) and the dish presentation was inspired in some of Massimo Bottura passions: traveling and art. Fig. 4 shows the dish inspired in a particular view of

Central Park in New York (USA), which represents the people (jelly base), the woods (foamed sauce) and the sky line with buildings (pieces of meat). A second version was also developed (Fig. 5) by settling the different six pieces of meat into a round dish. This new version kept the juice produced by the meat and mixed with the jelly base, which give a sweet and sour taste at the same time. Therefore, this last version would enhance the global flavor of the dish at a higher level. The order of eating the meat pieces in this second version dish must start from the head meat (*testina*) and eat the rest of the pieces clockwise (cheek, belly, tongue, tail and *cotechino*) to appreciate better the differences in flavor.

Figure. 3. Detail of peperonata sauces with yellow and red pepperoni.

This culinary concept clearly gives a step forward from a traditional product to an improved version from a nutritional point of view. These "new traditions" should give new hints to the normal food that people usually consume. Preliminary results have shown that this meat cooked by *sous vide* technique contains higher contents in multiple nutrients such as minerals than the meat cooked by boiling (Schellekens, 1996). The final and most important objective would be to change the traditional dish and incorporate this technique into the entire population. A new collaboration with the University of Parma (Italy) is carried out in order to integrate this technique in restaurants for workers to improve diet population. Therefore, working over the concept of linking tradition and innovation can give great benefits to society without forgetting the roots of our people.

Figure. 4. First version of *Bollito non-bollito* inspired in a particular view of Central Park in New York.

Figure. **5**. Final presentation of the second version dish *bollito non-bollito* (boiled non-boiled).

REFERENCES

1. Boni, Ada, 1971. Il talismano della felicita. Colombo Editorial. Rome, Italy.

2. Molinari Pradelli, Alessandro, 1999. Il grande libro della cucina italiana. Newton & Compton Editoral. Rome, Italy.

3. Artusi, P., 1891. Science in the kitchen and the art of eating well, The Lorenzo Da Ponte Italian Library. University of Toronto Press, Toronto, Canada

4. Baldwin,D.,2012.Sousvidecooking:areview.Int.J.Gastron.FoodSci.1 (1), 15–30.

5. Carnacina, Luigi, Veronelli, Luigi, 1960. La grande cucina. Garzanti Editorial. Milan, Italy.

6. Roca, J., Brugués, S., 2003. La cocina al vacío. Montagud Editores SA, Barcelona, Spain.

7. Schellekens,M.,1996.Newresearchissuesinsous-videcookingreview article. TrendsFoodSci.Technol.7(8),256–262.

8. Sultano, C., Bottura, M., 2006. Attraverso tradizione e innovazione. Bib- liotheca Culinaria, Rome, Italy.

6

THE ENGINEERING INSIDE OUR DISHES

Jose Miguel Aguilera

Department of Chemical and Bioprocess Engineering, Pontificia Universidad Católica de Chile, Avda. Vicuña Mackenna 4860, Macul, Santiago, Chile

ABSTRACT

For an engineer the real value of a product is not in its molecular composition but in the intrinsic properties derived from the structure that is formed. Nobody cares about the molecules in a cellular phone except that they have to be arranged to receive and emit calls in reliable form. In the case of foods this brings the focus to the "engineering inside the product" rather than on the process engineering of mixing, drying, heating, freezing and so on, which has been the traditional realm of food engineering.

The objective of this article is to introduce food scientists, chefs and amateur cooks to basic concepts and terminology used in food materials science, and to give examples of the engineering inside what we eat.

KEYWORDS

Cooking; Engineering; Taste; Microstructure

1. INTRODUCTION

The structures we eat as foods are derived directly from nature, transformed by processing or developed in the kitchen as recognizable meals and dishes. In fact, our senses are adapted to perceive and identify the unique properties of such structures in the form of appearance, texture, juiciness, sound, etc. Although the scientific study of cooking has traditionally concentrated on chemical and physico-chemical aspects, in the last decades much interest has arisen in understanding the formation and stability of food structures from the materials science viewpoint (Donald, 2004). Advances in this direction have been favored by the availability of powerful microscopes and analytical techniques that probe into the molecular mobility and localized mechanical and rheological properties.

By food microstructure we understand the spatial arrangement and interactions of identifiable elements in a food, whose sizes are <100 μm (Aguilera and Stanley, 1999).Fig. 1 shows several important structural elements related to foods, their approximate sizes as well as the sciences behind the phenomena at each length scale. The dimensions from molecules to products span almost eight decades. It is unfortunate that most of the structural engineering inside our foods occurs at sizes below 100 μm, being invisible to the naked eye.

For a physicist the range between the molecular size and the macroscopic scale is typical of soft condensed matter, or matter in a state between a liquid and a crystalline solid. At these dimensions molecules may participate in the formation of emulsions, viscous polymer solutions, gels and glasses. Thus, foods can be classed as soft matter but their multicomponent nature and complexity set them apart from other forms of soft matter present in our daily life (Mezzenga et al., 2005). One important characteristic of soft matter is that some molecules tend to self-assemble, a first step in the formation of food structures. Thus, some ingredients such as surfactants and globular proteins may spontaneously associate into micelles or gels, respectively, given the right conditions.

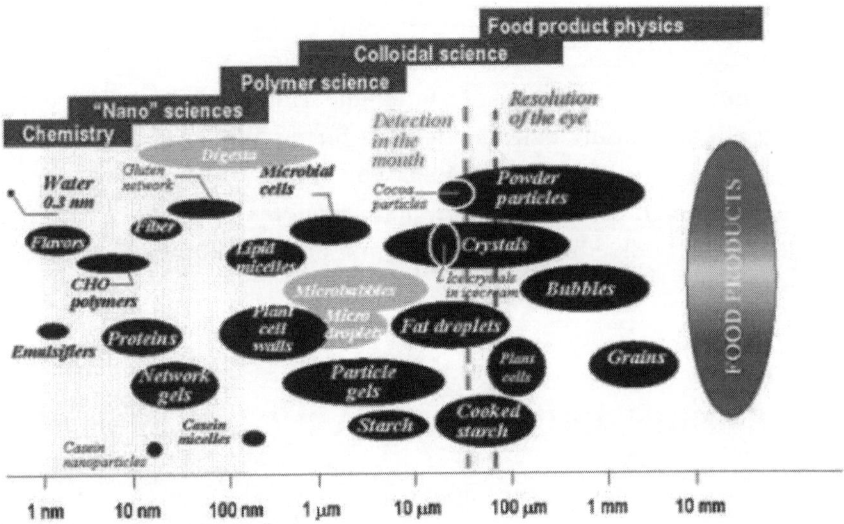

Figure. **1**. Important structural elements related to foods and their approximate sizes. Dotted lines show the upper limit for particles to go undetected in the mouth and the minimum size that can be resolved by the naked eye. Gray area is the size range of nanosciences.

For an engineer the real value of a product is not in its molecular composition but in the intrinsic properties derived from the structure that is formed. Nobody cares about the molecules in a cellular phone except that they have to be arranged to receive and emit calls in reliable form. In the case of foods this brings the focus to the "engineering inside the product" rather than on the process engineering of mixing, drying, heating, freezing and so on, which has been the traditional realm of food engineering. One example illustrating this point are liquid dairy cream and whipped cream. Although the initial composition is the same, whipped cream being a foam and having air (which is not considered as an ingredient) dispersed as bubbles is a product with a higher value when it comes to decorating cakes and desserts. There are many other examples where the composition of foods is not a good predictor of their properties, but only a source of molecules to be exploited in making delicious structures.

The objective of this article is to introduce food scientists, chefs and amateur cooks to basic concepts and terminology used in food materials science, and to give examples of the engineering inside what we eat [a glossary of gastronomy and engineering terms may be found in Alicia and elBullitaller (2006)]. The reader who wants to get serious on the subject of food materials science, its principles and applications, is referred to the book edited by Aguilera and Lillford (2008).

2. THE ENGINEERING THAT CANNOT BE SEEN BUT CAN BE TASTED

Interestingly, in reviewing the classic text on food and cooking by McGee (2004), the term structure is profusely used to describe the anatomical parts of organisms (that later will become foods) as well as the internal portions of processed foods as seen with an electron microscope. Engineering concepts such as stability of the structure of whipped cream (reinforced at the bubble interface by fat globules), the plasticity and elasticity of wheat dough, etc, are also referred to. This is a tacit recognition that a food "structure" is apparent to food scientists and cooks alike. People in referring to foods during mastication also use a terminology that is related to their structure: tough meat, grainy sauce, soft beans, etc.

In a first approach a food may be regarded as a building, a structure possessing many architectural elements made of different materials (e.g., glass windows, wooden doors, ceramic tiles, etc.) inserted within a continuous frame. Fresh plant foods derive most of their desirable properties from a structure formed by cells (around 100 μm in size) glued together by a complex pectin gel to make a tissue or organ. A thick wall, which is basically a polymer composite reinforced by cellulose fibers, surrounds each cell and provides resistance. Turgor, important in texture, is derived from the osmotic pressure exerted by the solutes in a solution that fills a vacuole inside the cell. Animal cells (e.g., muscle fibers), on the contrary, do not have walls and rely on an internal support system formed by proteins and encasing membranes to keep the cell's (in this case, fiber's) contents.

This construction analogy may be used to explain softening of tissues during cooking. Heating of grains and legumes swell the hydrated starch granules inside cells making them tender, but most of all, it solubilizes the cement binding cells together. Upon biting on a soft cooked bean individual cells slide one past the other, much in the same way as bricks stacked one on top of the other would fall in absence of mortar after being pushed. In a tough bean on the contrary, cells remain bound and mastication has to fracture the cell's content, which is equivalent to tearing down a wall made by bricks held by mortar. Meats also become tender by cooking when the collagen surrounding the fibers is solubilized.

Many processed foods may contain pores or air cells, crystalline and glassy phases, particles, oil droplets, etc, dispersed in a basic matrix (Fig. 2). The "architecture" in foods is represented by the size, number, distribution and interactions between these elements that are critical to the identity and properties of each food. Let us take lyophilized instant coffee as an example. A glassy matrix of polysaccharides encapsulates and protects the valuable volatile aromas, while interconnected pores permit rapid rehydration of particles in contact with hot water, solubilization of the matrix and the release of aromatic molecules. The size of surface pores also affects the color of the product as larger pores reflect less light making the coffee particle to look darker. In the case of chocolate, particles have to be ground to less than 40 μm in size (around half the diameter of hair) otherwise they would be detected in the mouth as "sandiness". In turn, cocoa butter has to crystallize in the right form, ("tempered") or the snap and gloss of a chocolate bar would be impaired.

But the analogy between foods and buildings, or for that matter, any other product of our daily life (watches, computers, cars, clothes, etc.) fails short of being always valid. Industrial products are *designed* and built using technological knowledge so they deliver the expected benefits. Their parts or elements are constructed from an ample array of materials and assembled piece by piece according to a precise sequence of steps dictated by blueprints and a Gantt chart. To build a car more than five thousand different parts have to go in specific places and interact in precise ways: some have to slide, other to rotate, several provide rigidity, but all of them must last for many years. In contrast, ice cream was never designed, its ingredients are limited and critical elements, such as ice crystals, develop in different sizes and shapes during freezing of the whole mass. Ice crystals have a finite life in the freezer and the structure of ice cream has to fail in the mouth to provide a creamy and cooling sensation while releasing flavors and aromas. A "designed" ice cream would probably start by making small and equally-sized ice crystals and then adding them to the rest of the components.

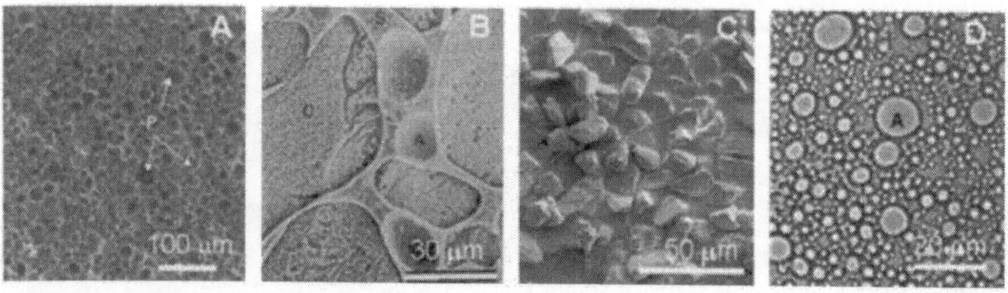

Figure. 2. Examples of the microstructures of some processed foods. (A) Freeze-dried instant coffee (P=pore); (B) ice cream (C=ice crystal, A=air, S=solution); (C) milk chocolate showing particles of cocoa, powder milk and sugar (courtesy of Dr. P. Braun, Buhler AG) and (D) mayonnaise (A=oil droplet).

It was only a few decades ago that food scientists started viewing food as structures and materials. This came about at realizing that food components – mainly water, proteins, carbohydrates and fats – by themselves could not explain the richness of textures and tastes of foods with similar composition. Although each component undergoes a series of changes during cooking (which are not different from those that they experiment in test tubes in the laboratory), it is in the interactions among them where most of the variability arises.

3. NANOTECHNOLOGY AND MICROTECHNOLOGY OF DAIRY PRODUCTS

It is remarkable that the wide assortment of dairy products is based principally on two building blocks: proteins (the casein micelles and globular whey proteins) and fat globules. In turn, casein micelles (ca. 300–400 nm in size) and fat globules (few μm in diameter) are a notable example of food nanotechnology as they are assembled inside the mammary cells of the cow's udder and released individually into the lumen where in conjunction with the globular whey proteins, lactose and minerals and other minor components form the milk. Thus, milk is as close to a "designed food" as we can think of: right nutrients in the form of versatile assemblies.

Emulsions (cream), foams (whipped cream), gels (yoghurt), a plastic solid (butter) and cheeses of many textures, among other dairy products, are the result of the interactions of these two types of elements (Fig. 3). Some interactions are driven by chemistry, as the gelling of casein micelles by the action of enzymes, while others are the result of surface phenomena (adsorption of whey proteins onto fat globules), or plain physics (grouping of fat globules denuded of membranes by mechanical action in butter formation).

4. MATERIALS SCIENCE IN THE KITCHEN

A most disregarded fact, but with important consequences, is that the foods we eat are rarely in equilibrium. A note before abandoning this section. Salt and table sugar are two ingredients in the crystalline (equilibrium) state and will remain like that for centuries if isolated from the environment. That foods are unstable is obvious for living tissues eaten raw, such as fruits and vegetables, which continue respiring after harvest and of meat and fish that suffer biochemical changes (not to mention microbiological decay). But approach to equilibrium is also apparent in foods that are served at temperatures above or below room temperature (e.g., hot plates and ice cream) that continuously exchange heat with the surroundings, thus, changing not only their temperature but the sensorial properties as well. Similarly, dry crunchy foods such as potato chips turn soggy and bread becomes dry by gaining and losing moisture to the neighboring atmosphere, respectively. More subtle are the cases of baked products becoming stale, food foams collapsing and emulsions separating into two phases upon storage. Notice that there is not major chemistry involved in the last examples, just a physical rearrangement of the components into a state of lower energy.

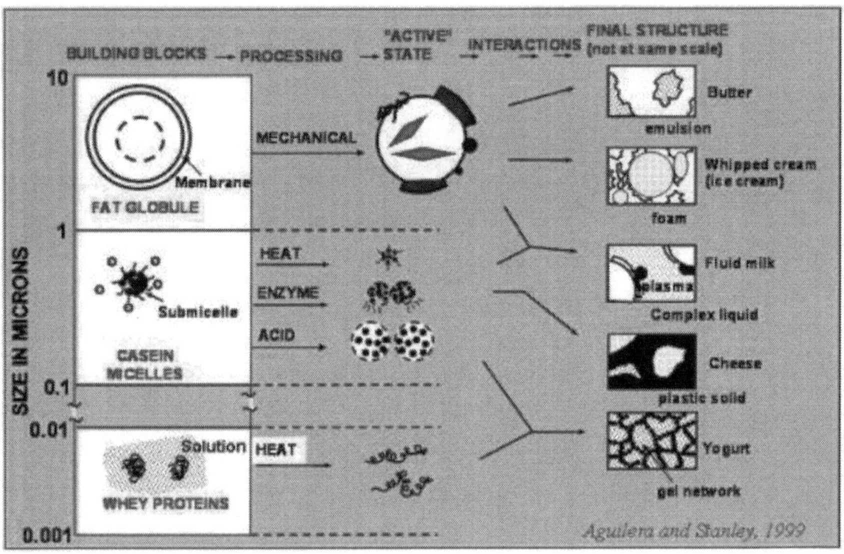

Figure. 3. The development of microstructures in dairy products. All is based on the interactions of 3 building units: casein micelles, whey proteins and fat globules.

Prolonging the stability of foods requires introducing a "barrier" opposing to the inevitable change. Process engineering makes use of external barriers such as a plastic foam package to isolate a warm food, or an impermeable plastic film to keep moisture out. Edible films have been suggested to separate wet and dry portions of a food (e.g., in sandwiches) thus retarding moisture migration. Of course, instability related to chemical kinetics can be controlled to a certain extent by lowering the temperature or modifying the atmosphere around the product.

But in the kitchen the physical stability may be extended by introducing the barrier opposing change "inside the product". In emulsions oil droplets with clean interfaces are subject to coalescence but large molecules or particles if positioned at their interfaces preclude their coming close together. Another alternative is to reduce the mobility of the system by increasing the viscosity of the continuous phase (e.g., adding a thickener). This and other cases require a good understanding of the role of thickeners, surfactants and other additives at the molecular and colloidal levels. For example, stabilizing an O/W emulsion requires a different type of surfactant than a W/O emulsion and in either case the amphiphilic molecule must migrate fast to the interface. Similar knowledge of the specific action of additives is required for the case of controlling fat blooming in chocolate, delaying staling of baked products, or avoiding caking of powders, among many other cases. Physics also dictates that a foam with uniform size bubbles (e.g., a monodispersed size distribution) is more stable than one having a mixed population of sizes. Frozen foods may appear to be stable in the domestic freezer (−18 °C) within the usual storage period between purchase and consumption, but they are not. Although separating water as ice reduces the reactivity of the system, the maximal stability is only attained when the remaining concentrated solution becomes vitrified (which in the case of fruits and vegetables occurs at around −40 °C).

Thus, producing metastable or unstable structures is inherent to cooking. The extreme case may be the soufflé, whose short life after removal from the oven is explained by the law of gases: as temperature is reduced, the pressure of air decreases and water vapor condenses, thus, the volume starts to diminish. On the contrary, hard candies made of amorphous sugar may be stable for years without becoming crystalline (the equilibrium state), if protected from humidity.

It is characteristic of foods that several structure-building phenomena, from the molecular (e.g., protein denaturation) to the macrostructural level (formation of air cells), occur more or less simultaneously. Thus, it is the relative kinetics of many events that drives structure formation in foods. Fig. 4 shows how temperature influences some of the transformations that major food components may undergo. As can be imagined, in a mixture of ingredients held at a certain temperature several phenomena may take place simultaneously. For instance, in the oven the proteins in a dough may become denatured, while starches undergo gelatinization and the water vapor released from the mass can be entrapped as air cells. Moreover, external transfer of heat originates temperature gradients within the product, thus, the rates of change will vary according to the position. At a certain point in time a structure sets in immobilizing the whole system. Besides temperature other variables may affect the kinetics of structure formation, among them, pH, ionic strength and the presence of certain ions. The microstructure of gels made of globular proteins varies depending on pH and ionic strength, hence, their properties: weak or strong, opaque or transparent, etc. A case in point is that of solutions of casein and whey proteins that may yield soft creamy gels or firm gels depending on the concentration of calcium.

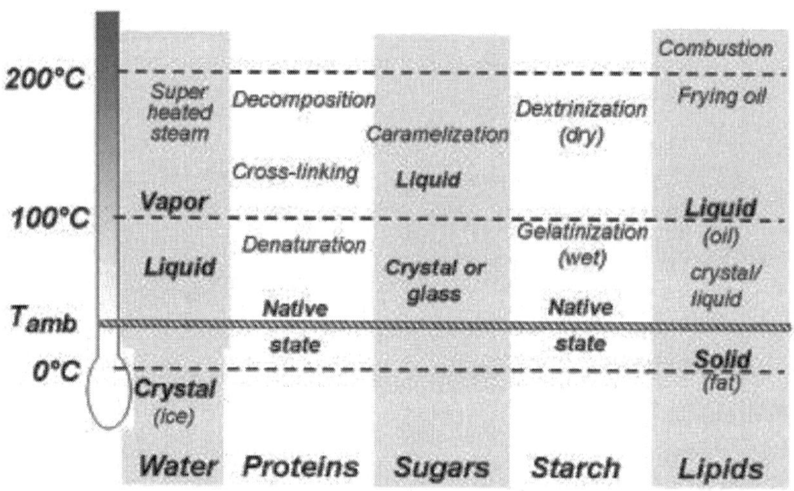

Figure. 4. Effect of temperature on phase changes, state transitions and some reactions of major components in foods. *Tamb*=ambient temperature.

It has been mentioned in the case of dairy products that the interaction between food components provides additional possibilities for structure development. Proteins and polysaccharides in solution may form association complexes through interactions of parts of the macromolecules having opposite charges. In addition, separation into a protein-rich phase and a polysaccharide-rich phase is not uncommon for concentrated solutions of this type of biopolymers. The disengagement of phases is generally slow, thus microstructures with different morphological characteristics may be "frozen" during the approach to equilibrium, for example, by gelation.

The state of "glass", taken from polymer physics, is the quintessential synonymous of a metastable condition for sugars, polysaccharides and some proteins in foods. Technically, a glass is formed when a pure liquid is cooled so rapidly that molecules are trapped as a solid occupying almost the same positions as in the liquid phase. Something similar occurs when water is removed from a sugar solution at a fast rate, as is the case of spray-dried products. Molecular mobility in the glassy state is significantly reduced, hence, food powders, crackers and pasta products exhibit excellent stability while they remain dry. Glassy foods have low moisture and are hard and brittle, emitting acoustic signals when fractured that are typical of crunchy foods. Unfortunately, glassy foods are avid for moisture and upon absorbing humidity from the atmosphere they become rubbery losing most of their desirable characteristics.

Salt and table sugar are two ingredients in the crystalline (equilibrium) state and will remain like that for centuries if isolated from the environment. But it is well known that some seeds remain viable (i.e., they can germinate) after being stored for long periods of time, meaning that they were "stable". Nature has engineered these seeds so that under low moisture conditions a glassy matrix made of sugars protects the genetic material and the enzymic machinery inside them.

5. WHAT IS AHEAD

In a sense, the processed food industry of the XX century was successful in providing low-cost, convenient products that continue to appeal to a large segment of the population. However, some of these foods (those so-called *junk food*) if consumed repeatedly may not be adequate for the present sedentary lifestyles and alternatives are urgently sought after to combat overweight and obesity. Here is an opportunity to design foods that are less calorie-dense than their highly demanded counterparts but equally tasty, since it is in human nature that what we eat has to be pleasant to our senses. Examples of improved products are most "light" versions of traditional foods. A second area where product design and engineering is likely to contribute is in the protection and target delivery of bioactive components and beneficial bacteria that have proven to be beneficial for our health.

At the other extreme are a growing minority of people around the world interested in eating well and experiencing novel sensations. They are open to new formats in cooking and enjoy the creativity and boldness of modern cooks. Some of the most reputed chefs are known to have their own laboratories or engage in associations with renowned scientists to bring innovation into their dishes. Edible films, cryogenic frozen desserts, foamed sauces, hot/cold gels, crunchy structures, artificial caviars, are just some of the proposals of the so-called *techno-emotive cuisine*. Again, there is a lot of engineering in these creations.

In academia, foods and gastronomy are becoming fashionable subjects that attract a multitude of students. An example of this trend is a new course at Harvard University bringing together famed chefs and eminent academics aiming to inspire students and advance kitchen science. In the words of David A. Weitz, Mallinckrodt Professor of Physics and of Applied Physics in the School of Engineering and Applied Sciences at Harvard "*much of what we do in the lab is what chefs like Ferran Adrià are now doing in their kitchens. Cooking provides an ideal framework to study a variety of complex phenomena—from basic chemistry to materials science to applied physics*".

As knowledge accumulates on the structure-forming capabilities of old and novel ingredients and how some of the complex microstructures of foods come into being, we will come closer to *designing* food structures for pleasure and health.

6. CONCLUSIONS

It is unfortunate that we cannot see the engineering inside our foods and dishes. It is like an architect being able to see only the outside of a house but not the layout of the rooms and the fine details of the decoration. In our mouths the engineering of foods is expressed in the breakdown of the food matrix, the release of flavors and juices and other multiple gratifying sensations. In our gut the structure of foods is likely to influence the bioavailability of many nutrients. For the first time in the history of foods and gastronomy a good body of scientific knowledge has accumulated that allows food technologists and chefs to design food structures for health and pleasure.

REFERENCES

1. Aguilera, J.M., Lillford, P.J., 2008. Food Materials Science: Principles and Practice. Springer, NY.

2. Aguilera, J.M., Stanley, D.W., 1999. Microstructural Principles of Food Processing and Engineering second ed. Aspen Publishers Inc., Gaitherburg, MD.

3. Alicia andel Bullitaller,2006.Le´xico Cientı´fico Gastrono´ mico. Editorial Planeta, Barcelona.

4. Donald, A., 2004. Food for thought. Nature Materials 3, 579–581. McGee, H., 2004. McGee on Food and Cooking: An Encyclopedia of Kitchen Science, History and Culture. Hodder and Stoughton Ltd., London.

5. Mezzenga, R., Schurterbenger, P., Burbridge, A., Michel, M., 2005. Understanding food as soft materials. Nature Materials 4, 729–740.

7

IMPROVEMENT OF BIOACTIVE COMPOUNDS CONTENT AND ANTIOXIDANT PROPERTIES IN CRACKERS WITH THE INCORPORATION OF PRICKLY PEAR AND POTATO PEELS POWDER

Yousif Elhassaneen, Sherif Ragab, Raghda Mashal

Nutrition and Food Science Department, Faculty of Home Economics, Minoufiya University, Shebin El-Kom, Egypt

ABSTRACT

Prickly pear peel (PPP) and potatoes peel (PP) are considered as a waste by-products which obtained during processing of prickly pear and potatoes, huge amount of peels are generated, and their disposal is a major problem and causes environmental pollution. In the present study, PPP and PP were dehydrated under vacuum at 70°C for 3 hrs to obtain prickly pear peel powder (PPPP) and potatoes peel powder (PPP) with 7% moisture content. These by-products were phytochemically analyzed, incorporated into crackers at 5% levels as a potential source of bioactive compounds. Bioactive compounds analysis indicated that PPPP and PPP contains high levels of many valuable bioactive compounds/antioxidants such total phenolics, (421 and 1388 mg GAE.g^{-1} DW) and carotenoids (217.11 and 135.76 mg.g^{-1} DW) as well as total dietary fiber (39.53 and 45.91 g.100g^{-1} DW). The total dietary fiber, carotenoids and total phenolics content in crackers increased from 5.89, 3.01 and 110.23 to 8.11, 14.34 and 143.28 with 5% incorporation of PPPP and 8.74 $g.100g^{-1}$, 7.88 $mg.g^{-1}$ and 192.79 mg $EGA.g^{-1}$ with 5% incorporation of PPP, respectively. Also, the antioxidant activity (AA) in control crackers was 30.11% which increased to 38.14 and 42.07% with the incorporation of PPPP and PPP by 5%, respectively. In conclusion, the results suggest that by incorporating PPPP and PPP, it is possible to enhance the nutritional and functional quality (bioactive compounds and antioxidant activity) of crackers without affecting on their sensory characteristics.

KEYWORDS

Prickly Pear, Potatoes, Peel, Powder, Dietary Fiber, Antioxidant Activity, Sensory Characteristics

1. INTRODUCTION

Agricultural and food industries in the Arab world represent a large proportion of waste was estimated at18.14 million tons per year and represent remnants of fruits and vegetables manufacture about 6.14% of this amount (http://elasaala. blogspot. com/2012/01/blog-post_2703.html). Processing of fruits, vegetables, and oilseeds may results in high amounts of waste materials such as peels, seeds, stones and oilseed meals. Disposal of these materials usually represents a problem that is further aggravated by legal restrictions. Plant waste is prone to microbial spoilage; therefore, drying is necessary before further exploitation. The cost of drying, storage, and transport poses additional economical limitations to waste utilization. Therefore, agro-industrial waste often is utilized as feed or fertilizer. However, demand for feed or fertilizer varies and depends on agricultural production. Moreover, valuable nutrients contained in agro-industrial wastes are lost. Thus new aspects concerning the use of these wastes as by-products for further exploitation on the production of food additives or supplements with high nutritional value have gained increasing interest because these are high-value products and their recovery may be economically attractive. It is well known that agro-industrial by-products are rich in dietary fibers, some of which contain appreciable amounts of colorants, antioxidant compounds or other substances with positive health effects, while some of them, like the oilseed meals, are rich in proteins [1]. Some major source of food wastes are potatoes and prickly pear, some of the most popular vegetables and fruits.

Potato (*Solanum tuberosum* L.) is the largest vegetable crop worldwide, amounting to approximately 320 million metric tons annually [2]. Processing of potatoes (mainly for the production of chips, French fries, and dehydrated products) has presented a steady increase during the last decades, exceeding considerably the amount of the vegetable consumed as fresh [1, 3, 4]. Solid waste generated during processing consists mostly of potato peels but also contains green, immature, and cull potatoes and amounts to 15-45% depending on the procedure applied [4]. It is used as animal feed, though fermentation for the production of single cell or alcohol has been considered [5]. Water from potato processing is used for the recovery of proteins by heat coagulation [5]. Recent investigations suggested the use of water extracts from potato processing waste for the recovery of antioxidants [6].

Prickly pear (*Opuntia ficus-indica*), commonly known as prickly pear, belongs to the family *Cactaceae*. Being so water-use efficient, they are highly useful in arid and semiarid environments, particularly during prolonged dry spells or failure of the monsoon [7]. So, it is widely distributed in Latin America, South Africa and the Mediterranean area including Egypt, [8]. It has been used in traditional folk medicine because of its role in treating a number of diseases and conditions, including anti-inflammatory effects [9], hypoglycemic effects [10], inhibition of stomach ulceration [11], neuroprotective effects [12]. Through antioxidant actions and also used for treating diabetes, hypertension, burns, bronchial, asthma edema, and indigestion in many countries over the world [13-14]. One of the most frequently utilized fruit and vegetable technologies is juice production. Juices, in general, are a good source of sugars, vitamins and minerals all valuable components to human health. In addition to their nutritional and medicinal properties, these plants contain compounds which have several commercial applications. In addition, the phenolic constituents of peels prickly pear cactus were isolated and identified using different chromatographic techniques [15]. Studies aims to magnify the using of prickly pear peel in food and nutritional applications are still a dearth, and more research is required.

Snack, defined as a light meal eaten between regular meals include a broad range of products that can take many forms. Definitions of snacks are being modified to include sandwiches, yogurt and even ice creams. Among snacks, crackers are the most popularly consumed bakery items parts of the world. Some of the reasons for such wide popularity are their ready to eat nature, affordable cost, good nutritional quality, availability in different tastes and longer shelf life [16]. Reported was available on the use of oat bran, wheat bran, rice bran as a source of dietary fiber content in bread and other bakery products [17-19]. Recently, Sudha *et al.*, [20] reported that influence of different cereal brans on sensory quality of biscuits. On the other hand, fruit dietary fiber concentrates have better nutritional quality than those found in cereals due to higher proportion of soluble dietary fiber (SDF) and significant content of dietary fiber associated bioactive compounds [21-22].

In Egypt, the bakery industry has grown tremendously over the recent years. Such as reported by Elhassaneen *et al.,* [23] and Ahmed, [24], bakery products are varied by addition of value added ingredients. Among the added ingredients, dietary fiber and phytochemicals have gained tremendous attention [25-26]. Several years ago, many studies reported the improvement in nutritional and functional properties of cookies and bread with the incorporation of flour with plant by-products powder [1, 4, 23, 26]. Although, there is an increasing demand for high fiber and phytochemicals ingredients in food products including snacks. Therefore, the objective of this work is to analyze some plant by-products including potatoes and prickly peach peels that constitute to waste for their nutraceutical values as food ingredient thus reducing its contribution to environmental pollution. Also, utilization of such by-products for their recovery of dietary fibers, bioactive compounds and antioxidants in the production of snacks such crackers will be in the scope of this study.

2. MATERIALS AND METHODS

2.1 Materials

2.1.1 Wheat Flour

Variety Giza 155 wheat (*Triticum vulgare*) was obtained from Shebin El-Kom market, Minoufiya Governorate, Egypt during the 2013 harvesting period. The collected samples was transported to the laboratory and stored immediately on the refrigerator at 0°C until using in preparation of flour. For flour preparation, the kernels was milled by using of laboratory mill (El-Araby, Benha, Egypt) and sieved through 60 and 50 meshes screen to obtain wheat flour extraction rate 72.

2.1.2 Food by-Products

Potato peel (*Solanum tuberosum* L.) was obtained from SFCO For Manufacturing & Export Agricultural Products, El Negila, Kom Hamada, Behira Government, Egypt; and prickly pear fruits (*Opuntia ficus-indica*), used for prickly pear peel preparation were obtained from the local markets of Tanta City, Egypt.

2.1.3 Crackers Ingredients

Salt, bicarbonate soda, yeast and shortening were purchased from the local markets of Shebin El-Kom City, Egypt.

2.2 Methods

2.2.1 Preparation of Food by-Products Peel Powder

(i). Prickly pear peel powder (PPPP): Unripe prickly pear pulp and unripe prickly pear peel were soaked in 0.1% sodium metabisulphite solution for 30 min, washed, sliced and dried at 60°C for 24 hours in hot air oven

(AFOS Mini Smoker, England). This is followed by milling with grinder (Retsch Micro Universal Bench Top Grinder, Germany) to produce the respective flour types.

(ii). *Potato peel powder (PPP):* Potato peels were washed and then dried in a hot air oven (Horizontal Forced Air Drier, Proctor and Schwartz Inc., Philadelphia, PA) at 55°C. The dried peels were ground into a fine powder in high mixer speed (Moulinex Egypt, El-Araby Co., Egypt). The material that passed through an 80 mesh sieve was retained for use.

2.2.2 Preparation of Crackers

The crackers were prepared according to the modified method of Manley, [27]. Formulation of the crackers is applied as follow: wheat flour, 100 g; shortening, 9.64 g; salt, 2.02 g; sodium bicarbonate, 0.16 g; yeast, 3.48 g and water 50 g. Yeast was mixed with water (25°C) to form a suspension, to which the other ingredients were then added and kneaded to form smooth dough. Substitution of wheat flour with prickly pear peel powder (PPPP) and potato peel powder (PPP) were conducted based on 5% of the weight of the wheat flour. The dough was later proofed for 2 hours in a proofer (Bakbar E81, New Zealand), followed by sheeting to 1.0mm thickness using a dough sheeter (Esmach, Italy). The dough was then cut into squares measuring 3cm x 3cm and 'docked' prior to baking at 170°C for 15 min.

2.2.3 Determination of Total Phenolics, Carotenoids, Total Dietary Fiber and Antioxidant Activity

Total phenolics, carotenoids and total dietary fiber in selected by-products and cracker samples were analyzed as follow: By-products and cracker samples were extracted with 80% acetone and centrifuged at 10,000g for 15 min. For cracker samples, one gram of biscuit powder was extracted with 20 ml of 80% acetone and centrifuged at 8000g at room temperature. The supernatant obtained from both samples were used for the analysis of total phenolics, carotenoids and antioxidant activity.

Total phenolics were determined using Folin-Ciocalteu reagent [28]. Two hundred milligrams of sample was extracted for 2 h with 2 mL of 80% MeOH containing 1% hydrochloric acid at room temperature on an orbital shaker set at 200 rpm. The mixture was centrifuged at 1000g for 15 min and the supernatant decanted into 4 mL vials. The pellets were combined and used for total phenolics assay. One hundred microliters of extract was mixed with 0.75 mL of Folin-Ciocalteu reagent (previously diluted 10-fold with distilled water) and allowed to stand at 22°C for 5 min; 0.75 ml of sodium bicarbonate (60g.L^{-1}) solution was added to the mixture after 90 min at 22°C, absorbance was measured at 725 nm. Results are expressed as gallic acid and equivalents. The total carotenoids in 80% acetone extract were determined by using the method reported by Litchenthaler, [29]. Total dietary fiber content in the MPP was estimated according to the method described by Asp *et al.,* [30].

Antioxidant activity of tested spices extracts and standards (a-tocopherol and BHT; Sigma Chemical Co., St. Louis, Mo) was determined according to the *b*-carotene bleaching method following a modification of the procedure described by Marco, [31]. For a typical assay, 1mL of *b*-carotene (Sigma) solution, 0.2 mg/mL in chloroform, was added to round-bottom flasks (50 mL) containing 0.02 mL of linoleic acid (J.T. Baker Chemical Co., Phillipsburg, NJ) and 0.2 mL of Tween 20 (BDH Chemical Co., Toronto, On). Each mixture was then dosed with 0.2 mL of 80% MeOH (as control) or corresponding plant extract or standard. After evaporation to dryness under vacuum at room temperature, oxygenated distilled water (50 ml) was added and the mixture was shaken to form a liposome solution. The samples were then subjected to thermal auto-oxidation at 50°C for 2 h. The absorbance of the solution at 470 nm was monitored on a spectrophotometer (Beckman DU-50) by taking measurements at 10 min intervals, and the rate of bleaching of *b*-carotene was calculated by fitting linear regression to data over time. All samples were assayed in triplicate. Various concentrations of BHT, BHA, and a-tocopherol in 80% methanol was used as the control. Antioxidant activity was calculated in four different ways. In the first, absorbance was plotted against time, as a knit curve, and the absolute value of slope was expressed as antioxidant value (AOX). Antioxidant activity (AA) was all calculated as percent inhibition relative to control using the equation (1) [32].

$$AA = (R_{control} - R_{sample}) / R_{control} \times 100 \quad (1)$$

Where: $R_{control}$ and R_{sample} were the bleaching rates of β-carotene in reactant mixture without antioxidant and with plant extract, respectively.

The third method of expression based on the oxidation rate ratio (ORR) was calculated according to the method of Marinova *et al.,* [33] using the equation (2):

$$ORR = R_{sample} / R_{control} \quad (2)$$

Where: $R_{control}$ and R_{sample} are the same in the previous equation.

In the fourth method, the antioxidant activity coefficient (AAC) was calculated as described by Mallet *et al.,* equation (3) [34].

$$AAC = (Abs_{S120} - Abs_{C120}) / Abs_{C0} - Abs_{C120}) \times 100 \quad (3)$$

Where: $Abs_{S\,120}$ was the absorbance of the antioxidant mixture at time 120 min, $Abs_{C\,120}$ was the absorbance of the control at time 120 min, $Abs_{C\,0}$was the absorbance of the control at zero time.

2.2.4 Sensory Evaluation

Sensory evaluation was carried out with 14 panelists comprising of postgraduate students from Minoufiya University, Shebin El-Kom, Egypt. Testing was done in the sensory laboratory. Each panelist was served with 5 randomly arranged cracker samples on a rectangular plastic tray. The crackers were individually sealed in a pouch and coded with a three-digit number prior to testing. The 5 samples consisted of two types of composite flour crackers and a control (100% wheat flour). Water was provided for rinsing between the samples. Panelists were required to evaluate the colour, crispiness, taste and overall acceptance of the crackers using the 9-point hedonic scale with 1=dislike extremely, 2=dislike very much, 3=dislike moderately, 4=dislike slightly, 5=neither like nor dislike, 6=like slightly, 7=like moderately, 8=like very much, and 9=like extremely.

2.2.5 Statistical Analysis

All measurements were done in triplicate and recorded as mean±SD. Statistical analysis was performed with the Student *t*-test and MINITAB 12 computer program (Minitab Inc., State College, PA).

3. RESULTS AND DISCUSSION

3.1 Analysis of Selected Food by-Products

3.1.1 Total Dietary Fiber, Carotenoids and Phenolics Contents

Total dietary fiber, carotenoids and phenolics contents of selected food by-products are shown in Table (1). The data showed that the total dietary fiber content was ranged 39.53 and 45.91 $g.100g^{-1}$, total carotenoids was 217.11 and 135.76 $mg.100g^{-1}$ and total phenolics was 421 and 1388 mg $GAE.100g^{-1}$ for PPPP and PPP, respectively. The PPP was recorded the highest content of total dietary fiber and total phenolics while PPPP recorded the highest values of total carotenoids. In similar study of Ahmed, [26] found that the total dietary fiber content of ROSP, MPP, CLP and PPP were ranged 27.15-42.71$g.100g^{-1}$, total carotenoids was 92.43-412.14 $mg.100g^{-1}$ and total phenolics was 1104-7129 mg EGA.100 g^{-1}. Such data confirmed that such tested by-product could be played a significant role in nutritional/pharmaceutical applications through their high content of bioactive compounds. Such as reviewed in Al-Weshahy and Rao, [35] dietary fiber is well known as a bulking agent, increasing the intestinal mobility and hydration of the feces. Several authors have reviewed the importance of consumption of moderate amounts of dietary fibers for human health [36-37]. Scientifically speaking, dietary fiber is a broad term that includes several carbohydrates; cellulose, hemicelluloses, lignins, pectins, gums etc. [38]. Camire *et al.,* [39] reported that PP fibers are primarily insoluble, and can bind bile acids *in-vitro*. It is believed that binding of bile acids is one of the mechanisms whereby certain sources of dietary fibers lower plasma cholesterol. Lazarov and Werman, [40] studied the hypocholesterolemic effect of dietary fiber from PP and found that after four weeks of feeding on potato peels, rats showed 40% reduction in plasma cholesterol content and 30% of hepatic fat cholesterol levels were reduced as compared with animals fed only with cellulose supplemented diet. Defects of dietary fiber on lipid-profile influence several health related issues. High concentrations of low-density lipoprotein (LDL) cholesterol, other dyslipidemia [high concentration of triglycerides and low concentration of high-density lipoprotein (HDL) cholesterol], leads to blood platelets aggregation [41], risk factors for cardiovascular diseases (CVD) [42], and hypertension [43]. Moreover, high intake of dietary fibers has a positive influence on blood glucose profile and it is related health complications, in healthy and diabetic individuals of both types. By altering the gastric emptying time, dietary fibers are able to affect the absorption of other simple sugars. The effect of dietary fibers on blood glucose and insulin response has been demonstrated by many other authors as well [35, 44]. Recently, Shalaby [25] reported that adding the food by-products such onion skin powder (high level of dietary fiber and phenolic compounds) to the ground meats before the grilling process and used in feeding of rats leads to decline in their serum glucose and lipid oxidation and improve in the liver functions.

3.1.2. Antioxidant Activities

The antioxidant activities and total phenolics of the tested food by-products are shown in Table (2). From such data it could be noticed that the by-products powder showed considerable differences in antioxidant activity (AA = 71.05 - 83.22%) when it was calculated by the four different methods used in this study. potato peels powder (PPP) showed strong activity because of its high phenolic content (1388 mg EGA.100g^{-1}DW) while prickly pear peel powder (PPPP) showed relatively low content in both antioxidant activity and the total phenolics (421 mg EGA.100g^{-1}). The decrease in absorbance of β-carotene in the presence of different methanolic tested food by-products extracts (and well-known antioxidants used as standards) with the oxidation of β-carotene and linoleic acid, antioxidant stability, is shown in Figure (1). Such data indicated that PPP recorded the lowest decreasing followed by PPPP. The values of PPP absorbance through 120 min are coming well i.e. corresponding to the line of 50 mg.L^{-1} of α-tocopherol and upper to the line 50 mg.L^{-1} of BHT. By the same manner, PPPPis coming close to the line of 50 mg.L^{-1} of BHT. These data proved the high stability of the all tested food by-products when comparing with that more common standards, BHT and α-tocopherol. In a similar study, Ahmed, [24] reported highly antioxidant activity (AA= 84.87-97.66%) and phenolic content (1679-8946 mg EGA.100g^{-1} DW) for many food by-products including PPP, red onion skin powder (ROSP) and mango peel powder (MPP). The high antioxidant stability (The decrease in absorbance of β-carotene in the presence of by-product methanolic extracts, and well-known antioxidants used as standards, with the oxidation of β-carotene and linoleic acid) was recorded for the methanol extract of ROSP followed by MPP and PPP, respectively.

Table 1. Total dietary fiber, carotenoids and phenolics contents of selected food by-products

Parameters	Prickly pear peel powder (PPPP)	Potato peel powder (PPP)
Total dietary fiber (g.100g^{-1})	39.53 ± 3.50	45.91 ± 3.83
Total carotenoids (mg.100g^{-1})	217.11 ± 20.44	135.76 ± 10.42
Total phenolics (mg EGA.100 g^{-1})	421 ± 64	1388 ± 98

[a] Each value represents the mean of three replicates ±SD.

Table 2. Antioxidant activity of selected food by-products

Samples	Antioxidant value[a] AOX (A/h)	Antioxidant activity[b] AA (%)	Oxidation rate ratio[c] (ORR)	Antioxidant activity coefficient[d] (AAC)
Prickly pear peel powder (PPP)	0.164 ± 0.021[a*]	71.05 ± 5.11[c]	0.289 ± 0.020[a]	407.48 ± 43.85[c]
Potatoes peel powder (PPP)	0.095 ± 0.011[b]	83.22 ± 2.12[b]	0.167 ± 0.011[b]	619.06 ± 50.87[b]
Control	0.565 ± 0.031	0.00 ± 0.00	0.998 ± 0.125	0.00 ± 0.00
BHT, 50 mg/L	0.079 ± 0.012[b]	86.03 ± 10.34[b]	0.139 ± 0.013[c]	667.91 ± 69.90[b]
α-tocopherol, 50 mg/L	0.011 ± 0.006[c]	97.99 ± 9.43[a]	0.020 ± 0.007[d]	875.83 ± 100.65[a]

[a]Antioxidant value (AOX, A/h) = The absolute value of slope (Abs was plotted against time).
[b]Antioxidant activity (AA,%) = (R control - R sample) / R control x 100 where: R control and R sample were the bleaching rates of beta-carotene in reactant mixture without antioxidant and with plant extract, respectively.
[c]Oxidation rate ratio (ORR) = R sample / R control.
[d]Antioxidant activity coefficient (AAC) = (Abs S 120 - Abs C 120) / Abs C 0 - Abs C 120) x 1000 where: Abs S 120 was the absorbance of the antioxidant mixture at time 120 min, Abs C 120 was the absorbance of the control at time 120 min, Abs C 0 was the absorbance of the control at zero time.
[*]Each value represents mean ±SD. Values with the different letters in the same column are significantly different $P \leq 0.05$.

Table 3. Total dietary fiber, carotenoids and phenolics contents and antioxidant activity of the control and composite flour crackers

Treatment	Total dietary fiber ($g.100g^{-1}$)	Total carotenoids ($mg.100g^{-1}$)	Total phenolics ($mg\ GAE.100g^{-1}$)	Antioxidant activity, AA (%)
Control crackers (CC)	5.89 ± 0.20^{b}	3.01 ± 0.62^{b}	110.23 ± 15.34^{c}	30.11 ± 2.17^{b}
CC + 5% PPPP	8.11 ± 0.32^{a}	14.34 ± 1.98^{a}	143.28 ± 9.81^{b}	38.14 ± 1.43^{a}
CC + 5% PPP	8.74 ± 0.24^{a}	7.88 ± 1.21^{b}	192.79 ± 21.67^{a}	42.07 ± 1.09^{a}

Each value represents the mean of ten replicates ±SD. Mean values with the different letters in the same column mean significantly different at level $p \leq 0.05$.

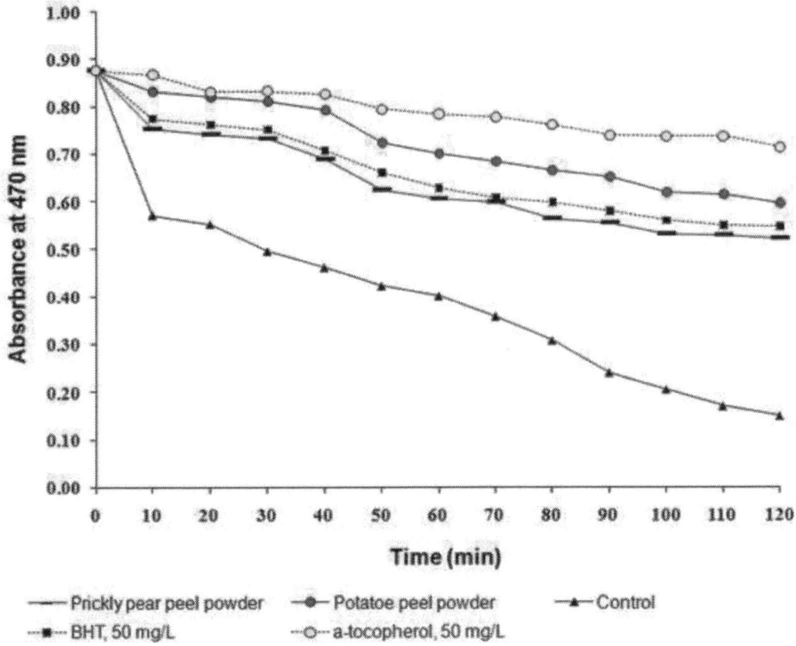

Figure 1. Activity of selected food by-products assayed by the b-carotene bleaching method (BHT a–tocopherol at 50 mg/L concentration was used as a reference).

In the correlation analysis, important differences were found between phenolics, carotenoids and antioxidant activity of food by-products (Figures 2). When all food by-products were included in the statistical analysis, there was a positive significant ($p \leq 0.05$) relationship between total phenolics in PPPP ($r^2 = 0.636$), total phenolics in PPP ($r^2 = 0.834$), total carotenoids in PPPP ($r^2 = 0.531$), total carotenoids in PPP ($r^2 = 0.593$) and antioxidant activity. These correlations confirm that phenolic compounds with carotenoids are mainly responsible for the antioxidant activity of the tested by-products. Also, these data indicates that many other bioactive compounds beside phenolics and carotenoids such vitamins, fibers, minerals etc probably contribute in the antioxidant activity of the tested by-products. This information was confirmed by Osuna-Martínez *et al.,* [45] who reported that antioxidant properties, one medicinal plant that has been proposed having interesting antioxidant activity and protective capacities due to the presence of components such as vitamins C and E, phenolics and other non-nutrient substances is *Opuntia ficus indica (L)*. In a similar study, Ahmed, [24] reported that positive and highly significant relationship between total phenolics and antioxidant activity for different food by-products including ROSP, PPP and MPP. Also, Velioglu *et al*., [46] reported that the correlation coefficient between total phenolics and antioxidative activities of 28 plant products and by-products, including sunflower seeds, flaxseeds, wheat germ, buckwheat, several fruits, vegetables, and medicinal plants was statistically significant. Recently, Abou-Elella and Ali [14] reported that the antioxidant of the Egyptian Prickly Pear Cactus (*Opuntia ficus-Indica*) Peel potency showed significant relationship with the total phenolic content.

3.2 Analysis of Crackers Enriched with Tested Plant by-Products

3.2.1 Bioactive Components and Antioxidant Activity of Control and Composite Flour Crackers

Table (3) shows the bioactive components and antioxidant activity of control and enriched crackers with the tested by-products. The total dietary fiber, carotenoids and phenolics content in crackers increased from 5.89, 3.01 and 110.23 to 8.11, 14.34 and 143.28 with 5% incorporation of PPPP and 8.74 $g.100g^{-1}$, 7.88 $mg.g^{-1}$ and 192.79 mg $EGA.g^{-1}$ with 5% incorporation of PPP, respectively. In parallel, the antioxidant activity (AA) in control crackers was 30.11% which increased to 38.14 and 42.07% with the incorporation of PPPP and PPP by 5%, respectively. Tested by-products enriched crackers showed strong activity probably due to their high bioactive compounds (carotenoids and phenolics) content.

Many similar studies indicated that big differentiations have been recorded amongst different tested by-products. For example, Onyeneho and Hettiarachchy, [47] reported that the peels from the red potatoes contained more polyphenols than those from the brown-skinned varieties. Furthermore it was shown that the peel and the pulp of the tubers contain nine phenolic acids differed in their concentrations and appeared to be mainly responsible for the strong antioxidant activities of the peel extracts. Ashoush and Gadallah, [48] investigated the effect of mango peels powders (MPP) at different replacing levels (5, 10, 15 and 20%) and mango kernels powders (MKP) at (20, 30, 40 and 50%) antioxidant properties of biscuits were evaluated. The content of phenolics increased from 3.84 to 24.37 mg/g of biscuit incorporated with deferent levels of MPP and MKP. The biscuits incorporated with MPP and MKP exhibited an improvement in their antioxidant properties.

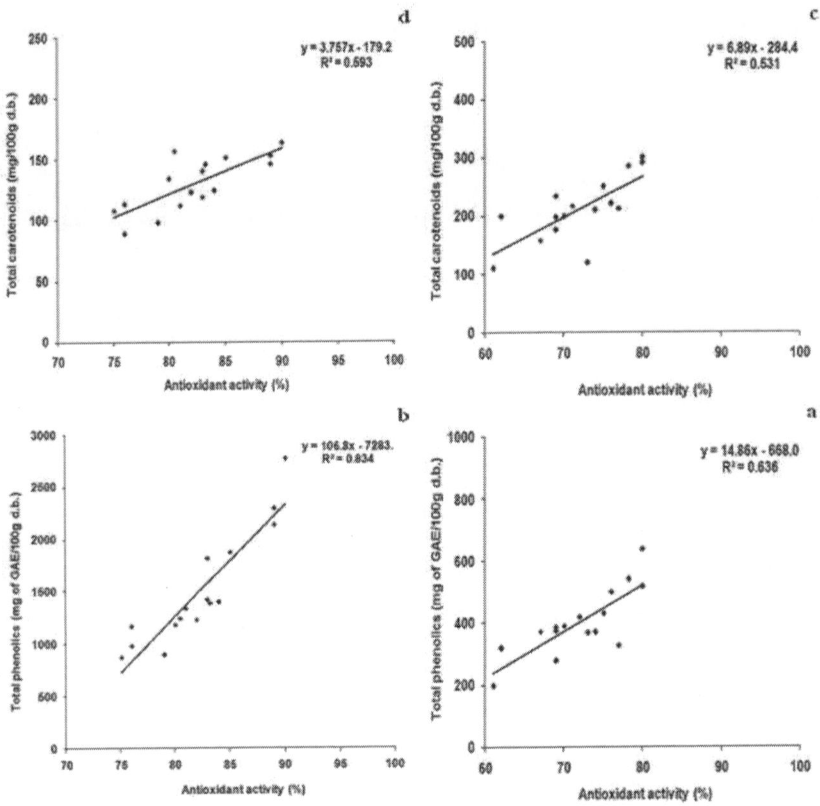

Figure 2. Correlation between bioactive components detected and antioxidant activity (AA) of both food by-products: (a) AA vs. total phenolics in PPPP, (b) AA vs. total phenolics in PPP, (c) AA vs. total carotenoids in PPPP, and (d) AA vs. total carotenoids in PPP.

Also, El-Sadany, [49] found that the mean of phenolic acids content was 2439.21± 113.8 mg/100 g extract of potato peel. Recently, Ahmed [26] indicated that breads formulated with food by-products (PPP, CLP, ROSP and MPP) exhibited increased total dietary fiber (TDF) content. The TDF content in control breads was 6.98 $g.100g^{-1}$which increased to 8.22, 8.72, 8.37 and 9.14 $g.100g^{-1}$ with the incorporation of PPP, CLP, ROSP and MPP by 5%. The total carotenoids content in control breads was 2.56 $mg.g^{-1}$ which increased to 8.90, 10.02, 7.01 and 22.04 $mg.g^{-1}$ with the incorporation of PPP, CLP, ROSP and MPP by 5%.The total phenolics content in control biscuits was 1.33 $mg.g^{-1}$which increased to 103.86, 56.12, 349.43 and 298.05 mg $EGA.g^{-1}$ with the incorporation of PPP, CLP, ROSP and MPP by 5%. Also, many studies carried out by Hegazy, [50], and Ahmed [26] indicated that there was a positive and significant ($p < 0.01$) relationship between total phenolics and antioxidant activity in different plant parts. Plant-based foods generally are considered important sources of antioxidants in the diet. Antioxidants help protect cells from the potentially damaging physiological process known as "oxidative stress" (damage to healthy cells or DNA by unpaired electrons known as free radicals). Oxidative stress is thought to be associated with the development of chronic diseases including cancer, heart disease, conditions of ageing including neurodegenerative diseases such as Parkinson's and Alzheimer's disease. There are a variety of plant antioxidants with different chemical structures. There are the antioxidant nutrients such as vitamins C, E, β-carotene and the trace element selenium (found in MPP) for which there are Dietary Reference Values (DRVs). However, there are thousands of other bioactive compounds in foods that have antioxidant activity but are not classified as "nutrients." These "non-nutrient antioxidants" include phenolic compounds such as found in PPPP and PPP [51].

3.2.2 Sensory Evaluation

Results of sensory evaluation in terms of colour, crunchiness, taste and overall acceptance are presented in Table (4). The PPPP and PPP crackers were significantly different ($p \leq 0.05$) in colour. The PPP crackers were shown to be the least liked by the panelists. It is reported by Broyart, [52] that the initial acceptance of baked products is much influenced by colour, which can also be an indicator of baking completion. As PPP flour imparts a dark brown colour to the crackers, this might have given the panelists an impression of 'over-baked' product, thus affecting their likings. For colour, among the tested food by-products, the colour of PPPP crackers was rated the highest (7.79) by the panelists. The desirable colour of crackers is mainly due to the Millard browning during baking. However, in PPPP crackers, the colour could be partially contributed by the carotenoid in PPPP flour which imparts a yellowish colour to the crackers. Similar findings were reported in a study by Brennan *et al.*, [53] who observed that an increased flour and thus muffin visual lightness (with more yellowness and brownness rather than dark and yellow green) yield a higher aroma, texture and colour acceptability scores. Colour was not significantly different ($p \leq 0.05$) between the control and PPPP crackers. Crispiness is perceived when food is chewed between molars, and is usually expressed in terms of hardness and fracturability. In this study, there was no significant difference ($p \leq 0.05$) in crispiness amongst the different samples with different composite flour cracker types. This observation could be due to the small percentage of wheat flour substitution in the crackers formulation, which did not affect the gluten network in the dough nor the development of an open internal structure upon baking.

In similar study, Sudha *et al.*, [20] reported that in biscuits prepared from different cereal fiber and found to be crispy at incorporation level of 40% oat bran, with a very small increase in hardness. No significant difference ($p \leq 0.05$) was observed in terms of taste between the control and PPPP crackers. This could probably be due to the nature of PPPP which did not impart any additional flavour to the crackers. However, PPPP was significantly different ($p \leq 0.05$) in taste as compared to the control and PPP crackers. The lowest score for the PPP crackers could be attributed to the 'potatony' after-taste, as commented by some of the panelists. There was no significant difference ($p \leq 0.05$) in term of overall acceptability among the control and PPPP. This could be attributed to the close resemblance of the cracker types in terms of the colour, crispiness and taste of the commercial crackers in the market. However, the PPP crackers were significantly different ($p \leq 0.05$) in overall acceptance. The lower ratings for the PPP crackers could be due to the unpleasant taste of the crackers. In similar study of Ashoush and Gadallah, [48] investigated the effect of mango peels powders (MPP) at different replacing levels (5, 10, 15 and 20%) and mango kernels powders (MKP) at (20, 30, 40 and 50%) on sensory properties of biscuits were evaluated. Acceptable biscuits with mango flavor were obtained by incorporating up to 10% MPP and with MKP up to 40%. Recently, Ahmed [26] acceptable breads with MPP and CLP were obtained by incorporating up to 5%. . Wheat flour was also substituted with PP in the production of white bread, but it increased crumb darkening and reduced the loaf volume [54]. Also, Orr *et al.,* [55] found that PP caused a musty odor in breads, but the extrusion of potato peel before its utilization can diminish this aroma from the final product. Additionally, Abdel-Magied, [56] mentioned the effect of using PP in biscuit processing. The resulted biscuits produced using 5 and 10% (w/w) of PP replaced wheat flour were smaller in terms of their stack weight and sensory score (color and appearance) in proportion to the levels used. In addition, supplementary biscuits were harder to bite than control biscuits.

4. CONCLUSION

The chemical composition of PPPP and PPP showed that they are a good source of dietary fibers and bioactive compounds such as carotenoids and phenolics. Crackers samples enriched with that by-products showed higher TDF, carotenoids and total phenolics content than the control samples. Increasing of such bioactive compounds in PPPP and PPP incorporated crackers exhibited improving of their antioxidant activity subsequently their properties as functional foods. The tested by-products incorporated crackers by 5% doesn't affect on their overall quality.

Table 4. Sensory evaluation (Scores) of the control and composite flour crackers

Treatment	Colour	Crunchiness	Taste	Overall acceptance
Control crackers (CC)	8.33 ± 0.21^a	8.01 ± 0.20^a	7.82 ± 0.14^a	8.22 ± 0.10^a
CC + 5% PPPP	7.79 ± 0.17^b	7.23 ± 0.14^a	7.09 ± 0.32^a	7.81 ± 0.29^a
CC + 5% PPP	6.51 ± 0.23^c	6.51 ± 0.20^a	6.01 ± 0.19^b	5.99 ± 0.15^b

[*] Each value represents the mean of ten replicates ±SD. Mean values with the different letters in the same column mean significantly different at level $p \leq 0.05$.

ACKNOWLEDGEMENTS

The financial support of Research Supporting Fund, Minoufiya University, Shebin El-Kom, Egypt is gratefully acknowledged. Sincere thanks are also extended to Dr. Zeinab Eissa for by-products samples preparation and lab assistance.

REFERENCES

1. Vasso, O. and Constantina, T. (2007). Utilization of Plant By-Products for the Recovery of Proteins, Dietary Fibers, Antioxidants, and Colorants. In: Utilization of By-Products and Treatment of Waste in the Food Industry. Ed. By: Vasso Oreopoulou and Winfried Russ. Springer ScienceBusiness Media, LLC, PP. 209-232.

2. FAO, (2005). Statistical Database athttp://apps.fao.org. Agriculture data: Agricultural production-Crops Primary.

3. Kadam, S.; Natu, R.; Mazza, G. and Jadhav, S. (1991). Waste utilization. In: Salunkhe DK, Jadhav SJ, editors. Potato: production, processing, and products: CRC Press, Boca Raton, USA, PP. 175-199.

4. Schieber, A.; Stintzing, F. and Carle, R. (2001). By-products of plant food processing as a source of functional compounds-recent developments. Trends in Food Science and Technology, 12(11): 401-413.

5. Natu, R.; Mazza, G.; Jadhav, S. and Waste, U. (1991). In: Salunkhe DK, Kadam SS, Jadhav SJ, editors. Potato: production, processing, and products: CRC Press, Boca Raton, U.S.A. 175-199.

6. Rodriguez De Sotillo, D.; Hadley, M. and Holm, E. (1994). Potato peel waste: stability and antioxidant activity of a freeze-dried extract. Journal of Food Science, 59: 1031–1033.

7. Sing, G. (2003).General Review of Opuntias in India. Journal of the Professional Association for Cactus Development.

8. Hassan, A. (2011). The effect of phytochemicals on prevention and/or treatment of liver cancer induced by some food pollutants. Ph.D. Thesis, Faculty of Home Economics, Minoufiya University, Shebin El-Kom, Egypt.

9. Park, E.; Kahng, J. and Paek, E. (1998). Studies on the pharmacological actions of cactus: identification of is anti-inflammatory effect. Arch. Pharm. Res, 21: 30–34.

10. Frati, A.; Gordillo, B.; Altamirano, P.; Ariza, R.; Cortes-Franco, R. and Chavez, N. (1990).Acute hypoglycemic effect of *Opuntia steptacantha*Lemaire in NIDDM. Diabetes Care, 13: 455-456.

11. Galati, E.; Mondello, M.; Giufferida, D.; Dugo, G.; Miceli, N.; Pergolizzi, S. and Taviano, M. (2003).Chemical characterization and biological effects of Sicilian *Opuntia ficus indica* (L.) Mill. Fruit juice: antioxidant and antiulcerogenic activity. J. Agric. Food Chem., 51: 4903–4908.

12. Dok-Go, H.; Lee, K.; Kim, H.; Lee, E.; Song, J.; Lee, Y. and Jin, C. (2003). Neuroprotective effects of antioxidative flavonoid, quercetin, (1) dihydroquercetin and quercetin 3-methyl ether, isolated from *Opuntia ficus indica*var. saboten. Brain Research, 965: 130–136.

13. Cardador-Martínez, A.; Jiménez-Martínez, C. and Sandoval, G. (2011). Revalorization of cactus pear (*Opuntia spp.*) wastes as a source of antioxidants. Ciê e Tecn.Alim, 31: 782-788.

14. Abou-Elella, F. and Ali, R. (2014). Antioxidant and Anticancer Activities of Different Constituents Extracted from Egyptian Prickly Pear Cactus (*Opuntia ficus Indica*) Peel.Biochem Anal Biochem, 3: 158.

15. Huseain, A. E. (2013). Technological studies on some vegetables and fruits peels and their utilization on rasing the nutritional values of child foods" Ph.D. Thesis in Nutrition and Food Science, Faculty of Home Economics, Minoufiya University, Egypt.

16. Gandhi, A.; Kotawaliwale, N.; Kawalkar, J.; Srivastava, D.; Parihar, V. and Raghu, P. (2001). Effect of incorporation of defatted soy flour on the quality of sweet biscuits. Journal of Food Science and Technology, 38(5): 502-503.

17. Laurikainen, T.; Harkonen, H.; Autio, K. and Poutanen, K. (1998).Effects of enzymes in fiber-enriched baking. J. Sci. Food Agr, 76: 239-249.

18. Leelavathi, K. and HaridasRao, P. (1993).Development of high fiber biscuits using wheat bran. Journal of Food Science and Technology, 30(3): 187-191.

19. Saunders, R. (1990).The properties of rice bran as foodstuff. Cereal Foods World, 35(7): 632-634.

20. Sudha, M.; Vetrimani, R. and Leelavathi, K. (2007). Influence of fibre from different cereals on the rheological characteristics of wheat flour dough and on biscuit quality. Food Chemistry, 100 (4): 1365–1370.

21. Grigelmo-Miguel, N. and Martin-Belloso, O. (1999). Comparison of dietary fiber from by-products of processing fruits and greens and from cereals. Lebensmittel-Wissenschaft– Food Science and Technology, 32(8): 503-508.

22. Chau, C. and Huang, Y. (2003). Comparison of the chemical composition and physicochemical properties of different fibers prepared from the peel of Citrus sinensis L. Cv. Liucheng. Journal of Agricultural and Food Chemistry, 51 (9): 2615-2618.

23. Elhassaneen, Y.; Sherif, R.; Alaa, E. and Abeer, E. (2013). Mango peel powder: A potential source of phenolics, carotenoids and dietary fiber in Biscuits preparations 2nd International-16th Arab Conference of Home Economics "Home Economics in the Service of Industry"10-11 September, Faculty of Home Economics, Minoufiya University, Shebin El-Kom, Egypt.

24. Ahmed, S.K. (2015). Utilization of by-products of food industries in the production of snacks with high nutritional value and healthy safe " Ph.D. Thesis in Nutrition and Food Science, Faculty of Home Economics, Minoufiya University, Egypt.

25. Shalaby, H. H. (2015). The effect of some food products mixed with plant parts on blood sugar levels of rats " Ph.D. Thesis in Nutrition and Food Science, Faculty of Home Economics, Minoufiya University, Egypt.

26. Ahmed, A. S. (2016). Nutritional and technological studies on the effect ofphytochemicals on obesity injuries and their related diseases by using experimental animals ".Ph.D. Thesis in Nutrition and Food Science, Faculty of Specific Education, Port Saied University, Port Saied, Egypt.

27. Manley, D. (2001). Biscuit, cracker and cookie recipes for the food industry. Cambridge:Woodhead Publishers.

28. Singleton, V. and Rossi, J. (1965). Colorimetry of total phenolics with phosphomolybdic-phosphotungstic acid reagents. Am. J. Enol. Vitic., 16: 144-158.

29. Litchenthaler, H. K. (1987). Chlorophylls and carotenoids: pigments of photosynthetic biomembranes. Methods in Enzymology, 148: 350-383.

30. Asp, N. G.; Johnson, C. G.; Hallomer, H. and Siljestrom, H. (1983). Rapid enzymatic assay of insoluble dietary fiber.*Journal of Agricultural and Food Chemistry*, 31: 476-482.

31. Marco, G. (1968). A rapid method for evaluation of antioxidants. J. Am. Oil Chem. Soc., 45: 594-598.

32. Al-Saikhan, M.; Howard, L. and Miller, J. (1995). Antioxidant activity and total phenolics in different genotypes of potato (*Solanum tuberosum*, L.). J. Food Sci., 60(2): 341-343.

33. Marinova, E.; Yanishlieva, N. and Kostova, I. (1994). Antioxidative action of the ethanolic extract and some hydroxycoumarins of Fraxinusornus bark. Food Chem., 51: 125-132.

34. Mallett, J.; Cerrati, C.; Ucciani, E.; Gamisana, J. and Gruber, M. (1994). Antioxidant activity of plant leaves in relation to their α-tocopherol content.Jornal of Food Chem., 49: 61-65.

35. Al-Weshahy, A. and Rao, V. (2012). Potato Peel as a Source of Important Phytochemical Antioxidant Nutraceuticals and Their Role in Human Health - A Review, Phytochemicals as Nutraceuticals – Global Approaches to Their Role in Nutrition and Health, Dr Venketeshwer Rao (Ed.), InTech, Rijeka, Croatia.

36. Forsythe, W.; Chenoweth, W. and Bennink, M. (1976).The effect of various dietary fibers on serum cholesterol and laxation in the rat. Journal of Nutrition, 106: 26-32.

37. Ballesteros, M.; Cabrera, R.; Saucedo, M.; Yepiz-Plascencia, G.; Ortega, M. and Valencia, M. (2001). Dietary fiber and lifestyle influence serum lipids in free living adult men. Journal of the American College of Nutrition, 20(6): 649-655.

38. Gallaher, D. and Schneeman, B. (2001). Dietary fibre. In: Bowman AB, Russell MR, editors. Present knowledge in nutrition. Washington, DC, ILSI, 805.

39. Camire, M.; Zhao, J. and Violette, D. (1993). *In vitro* binding of bile acids by extruded potato peels. Journal of Agricultural and Food Chemistry, 41(12): 2391-2394.

40. Lazarov, K. and Werman, M. (1996). Hypocholesterlaemic effect of potato peels as a dietary fiber source. Journal of Medical Sciences Research, 24: 581-582.

41. Bagger, M.; Andersen, O.; Nielsen, J. and Ryttig, K. (1996). Dietary fibres reduce blood pressure, serum total cholesterol and platelet aggregation in rats. British Journal of Nutrition, 75(3): 483-493.

42. Erkkila, A. and Lichtenstein, A. (2006). Fiber and cardiovascular disease risk: how strong is the evidence? Journal of Cardiovascular Nursing, 21(1): 3-8.

43. Alonso, A., Beunza, J., Bes-Rastrollo, M., Pajares, R., Martinez, M. (2006). Vegetable protein and fiber from cereal are inversely associated with the risk of hypertension in a Spanish cohort. Archives of Medical Research, 37(6): 778-786.

44. Chandalia, M.; Garg, A.; Lutjohann, D.; von Bergmann, k.; Grundy, S. and Brinkley, L. (2000). Beneficial effect of high dietary fiber intake in patients with type 2 diabetes mellitus. The New England Journal of Medicine, 342: 1392-1398.

45. Osuna-Martinez, U.; Reyes-Esparza, J. and Rodríguez-Fragoso, L. (2014). Cactus (*Opuntia ficus indica*): A Review on its Antioxidants Properties and Potential Pharmacological Use in Chronic Diseases. University of Sinaloa, Faculty of Biological and Chemical Sciences, Cualiacan, Mexico. Nat Prod Chem Res, 2:153-161.

46. Velioglu, Y. S.; Mazza, G.; Gao, L. and Oomah, B. D. (1998). Antioxidant activity and total phenolics in selected fruits, vegetables and grain products. J. Agric. Food Chem., 46 (10): 4113-4117.

47. Onyeneho, N. and Hettiarachchy, S. (1993). Antioxidant activity, fatty acids and phenolic acids compositions of potato peels. Journal of the Science of Food and Agriculture, 62: 345-350.

48. Ashoush, I. and Gadallah, E. (2011). Utilization of mango peels and seed kernels powders as sources of phytochemicals in Biscuit. World Journal of Dairy & Food Sciences, 6 (1): 35-42.

49. El-Sadany, M. (2001). The effect of dietary phytochemicals on the prevention of liver cancer initiation induced by some chemical carcinogenesis. M.Sc. Thesis, Fac. of Home Economics, Minufiya University, Shebin El-Kom, Egypt.

50. Hegazy, W. (2009). Antioxidant activity of pomegranate (*Punica granatum*) fruits and its relationship with phenolic composition and

processing technology. M.Sc. Thesis in Nutrition and Food Science, Faculty of Home Economics, Minoufiya University, Egypt.

51. Ajila, C.; Leelavathi, K. and Prasada, J. (2008). Improvement of dietary fiber content and antioxidant properties in soft dough biscuit with the incorporation of mango peel powder. J. Cereal Sci., 48: 319-326.

52. Broyart, B. (1998). Predicting colour kinetics during cracker backing. J Food Eng, 35(3): 351-368

53. Brennan, J.; Norris, D.; Rodriguez, T.; Beddington, R. and Robertson, E. (2001). Nodal signaling in the epiblast patterns the early mouse embryo. Nature, 411: 965-969.

54. Toma, R.; Orr, P.; Appolonia, B.; Dintzis, F. and Tabekhia, M. (1979). Physical and chemical properties of potato peel as a source of dietary fiber in bread. Journal of Food Science, 44(5): 1403–1407.

55. Orr, P.; Toma, R.; Munson, S. and D'Appolonia, B. (1982). Sensory evaluation of breads containing various levels of potato peel. American Potato Journal, 59(12): 605-612.

56. Abdel-Magied, M. (1991). Effect of dietary fiber of potato peel on the rheological and organoleptic characteristics of biscuits. Egyptian Journal of Food Science, 19:293-300.

8

THE ROLE OF MEAT IN A HEALTHFUL DIETARY PATTERN: EVIDENCE FROM MENU MODELING

Betsy Hornick[1*] and Neva Cochran[2]

[1]Food, Nutrition, and Health Communications, Poplar Grove, IL 61065, USA

[2]Nutrition Communications Consultant, Dallas, TX 75230, USA

1. INTRODUCTION

Meat has comprised a substantial proportion of the diet dating back to early civilizations, serving as an important source of energy and nutrients while providing taste and satisfaction. Today, meat is an integral part of our food culture, contributing high quality protein and essential nutrients, including iron, zinc, B vitamins, selenium, choline, and potassium [1]. Consumer research suggests that almost all Americans eat meat; according to a 2012 Gallup Poll, about 5% of American adults (n=1,014) consider themselves vegetarian, largely unchanged from 6% in 1999 and 2001 [2]. Intake data from What We Eat in America (WWEIA) 2011-2012 indicates that Americans 2 years and over consume on average 3.9 ounces (oz) of meat, poultry and cured meat per day, representing 69% of total Protein Food Group intake, and ranging from 3.1 oz per day for adult females to 5.2 oz for adult males (Table 1) [3]. These intake levels fall within the 2010 Dietary Guidelines for Americans (DGA) US Department of Agriculture

(USDA) Food Pattern recommended daily intake of 5.5 ounce equivalents (oz eq) of Protein Foods for a 2,000-calorie diet [4].

The recent Scientific Report of the 2015 Dietary Guidelines Advisory Committee (DGAC) described dietary patterns associated with positive health outcomes for the US population as "lower in red and processed meat," with little recognition of the role of lean meat as an important source of protein, iron, and other essential nutrients [5]. However, recommended daily meat servings in the proposed DGAC Healthy US.-Style Pattern and Healthy Mediterranean-Style Pattern are unchanged from the 2010 DGA, suggesting there is no target reduction for red and processed meats. Two USDA Food Patterns (Healthy US.Style and Healthy Mediterranean-Style) include 12.5 oz eq per week (1.8 oz eq per day) of meat and 10.5 oz eq per week (1.5 oz eq per day) of poultry. Both categories include fresh and processed meats according to USDA Item Clusters and Representative Foods [5].

Varied choices of fresh and processed meats can serve as anchors in a healthful diet. Pairing of meat with whole grains, vegetables, fruits, and low-fat or non-fat dairy may help to increase servings of these under consumed food groups. Processed meats include meat and poultry products that have been changed from their original state by at least one processing or preparation step (Table 2). The purpose of this paper is to demonstrate that a variety of meats can be part of a healthful dietary pattern and play an important role in meeting nutrient needs for all ages. USDA Healthy Food Patterns, as outlined in the 2010 DGA and in the 2015 DGAC Report reinforce the role of meat, poultry, and seafood as part of a healthy dietary pattern rich in vegetables, fruits, whole grains, and dairy. Evidence-based dietary patterns identified in the 2015 DGAC Report suggest that there is more than one approach to consuming a healthful diet, with introduction of a Healthy Mediterranean-Style Pattern based on evidence that Mediterraneanstyle diets with higher amounts of fruit, seafood and red/processed meats offer health benefits [5]. Menu models (Table 3) that illustrate how real foods fit into USDA/MyPlate food patterns, demonstrate that meat and poultry, including red and processed meats, can meet current energy and nutrient recommendations, including nutrients of concern, and fall within limits for over consumed nutrients (saturated fat, sodium). The menu models provide realistic examples that show how consumers can achieve healthy dietary patterns with a variety of foods.

2. MEAT AND PROTEIN FOOD RECOMMENDATIONS AND CONSUMPTION

Since their introduction in the early 1900s, all USDA Food Guides have included a group for meat and protein foods. Although serving guidance has evolved, lean meat was recognized as early as 1930 in food group recommendations [6]. Consumer messaging to "Go lean with protein," first introduced in the 2005 DGA, has resonated with Americans, promoting greater awareness of lean meat in a healthful diet as well as greater availability of lean meat options [7]. Greater availability of lean meat and processed meat in portion-controlled packaging and reduced fat or reduced sodium options make it possible to include meats in a healthful eating pattern consistent with public health recommendations [8]. The 2010 DGA emphasize "high quality protein" as the preferred source, but provide little guidance on specific criteria to categorize protein quality [4], which is determined by its essential amino acid (EAA) content, unique amino acid profile, and the digestibility of each EAA in the protein [9].

Dietary guidance, beginning with the release of the first edition of the Dietary Guidelines for Americans in 1980, has consistently recommended consumption of meat that is lean. This has been reinforced by enduring guidance to choose diets lower in total fat, saturated fat, cholesterol, and sodium. In 2000, guidance for total fat was modified to choosing a diet moderate in total fat. The meat industry has responded to the demand for leaner meats with improved breeding and feeding practices, and advanced trimming and processing to help lower fat and sodium content. For example, 66% of beef cuts sold at retail qualify as "lean" (when cooked and trimmed), and sirloin steak now contains 34% less fat than it did in the 1960s [7]. Nutrition education and advice to choose lean meat has been shown to be effective with shoppers, putting greater emphasis on leaner cuts and portion control [8].

Table 1: Current Intakes, Contributions, and Recommendations for Meat, Poultry, and Cured Meats.

	Current Intakes, oz/d (WWEIA 2001-2012)			% Contribution from WWEIA Food Categories (2009-2010)			Food Pattern Recommendation
	Ages 2+	Males 20+	Females 20+	Calories	Saturated Fat	Sodium	2015 DGAC (2000 calorie)
Meat	1.56	2.26	1.19	2.12%	3.33%	2.10%	12.5 oz/wk (1.8 oz/d)
Poultry	1.37	1.71	1.18	3.31%	3.33%	4.41%	10.5 oz/wk (1.5 oz/d)
Cured Meat	0.98	1.27	0.74	1.32%	2.85%	3.18%	--
Total Meat, Poultry, Cured Meat	3.9	5.24	3.11	6.75%	9.50%	9.69%	23 oz/wk ((3.3 oz/d)
Total Protein Foods	5.7	7.59	4.82	11%	15.03%	14.06%	5.5 oz/d

Table 2: Types of Meats Defined.

Lean meat	Defined by the USDA, a serving of meat qualifies as "lean" if it contains 4.5 grams or less of saturated fat, less than 10 grams of total fat, and less than 95 milligrams of cholesterol per 100 grams (3.5 ounces) [54]
Processed meat	A broad category of meat and poultry products that have undergone at least one further processing or preparation step that changes the appearance, texture, or taste, such as grinding, adding an ingredient, or cooking. Processed meats may be ready-to-cook or ready-to-eat. Variations in formulations can allow for lean or low-fat and low sodium products, among other choices.
Red meat	No universal definition exists, but generally includes meat from livestock (beef, veal, pork, and lamb) that contain higher levels of myoglobin, a protein that holds oxygen in the muscle and gives meat its reddish color. Red meat may be fresh or processed.
Cured meat	Meat pieces are treated with small amounts of nitrite, either as dry salt or as salt solution in water, to impart a characteristic pink color and distinct taste. Cured meat is primarily consumed as cooked meat, but certain regional products may be raw.
Smoked meat	Meat is flavored, cooked, and preserved by exposing it to smoke from burning or smoldering material, most often wood, or through application of liquid smoke during thermal processing.
Dried meat	Dried meat products are the result of the simple drying of lean meat in natural conditions or in an artificially created environment to meet specific water activity levels.

Myriad changes over the last century have influenced meat consumption trends, including rising incomes and trade, improved understanding of nutrition and health, new technologies and product development, agricultural policies, and changing social norms [10]. Since the early 1900s, when food availability data was first collected, overall meat consumption has increased [11]. However, USDA data suggests that meat consumption has peaked with recent lossadjusted food availability data indicating that per capita annual meat consumption fell from a high of 146 pounds in 2004 to 132 pounds in 2012 [12]. From 1970 to 2012, an overall downward trend included a 26% reduction in red meat consumption (beef, veal, pork, and lamb), decreasing from 4.2 oz to 3.1 oz per day, with doubled intakes of poultry, increasing from 1.2 oz to 2.4 oz per day [12]. Nutrient availability data, including protein, iron, zinc, and B vitamins, per capita per day in meat and poultry available for consumption over the past decade, and the percentage contributions of these nutrients from meat and poultry, are consistent with these trends [13].

The 2015 DGAC Report recognizes that lean meats can be part of a healthy dietary pattern as shown by USDA Healthy Food Pattern modeling [5]. Consumption data indicates that Americans are currently consuming amounts of meat consistent with the 2010 Dietary Guidelines and the proposed 2015 DGAC Food Patterns, which are based on intake data from WWEIA 2009-2010 (Table 1). According to WWEIA 2011-2012, average intakes of meat, poultry, and cured meats remain within the daily recommended intake of Protein Foods of 5.5 oz/day, which also includes seafood, eggs, nuts and seeds, and are comparable to the proposed USDA Healthy US-Style Food Pattern for 2,000 calories at 26 oz per week (3.7 oz/day) of meat, poultry, and eggs. For adult females 20 years and over, total intake of meat, poultry, and cured meat is below the 2015 DGAC Healthy US.-Style Food Patterns recommendation of 23-28 oz/week (3.3-4 oz/day) proposed for calorie levels ranging from 1,600-2,200; and for adult males, intakes are proportionate to higher energy needs and slightly above the allowance of 31-33 oz/week (4.4-4.7 oz/day) proposed for 2,400-3,000 calories [5].

According to the 2015 DGAC Report, meat, poultry, and cured meats together contribute 6.75% of energy, 9.5% of saturated fat, and 9.7% of sodium intakes in the US population ages 2 and older (data derived from WWEIA 2009-2010) [5]. However, this does not include meat and poultry consumed in burgers, tacos, sandwiches, and mixed dishes, which contribute heavily to intakes of energy, saturated fat, and sodium, but also contribute other food groups and nutrients including grains, vegetables, and dairy foods, as well as under consumed nutrients (fiber, calcium, potassium, and vitamin D). Total beef consumption contributes 5% of the calories and 10% or less of total fat and saturated fat in the US diet [1]. Consumption of lean pork contributes 3% of total energy, fat, and saturated fat, and 10% of total protein intake in the diets of lean pork consumers [14]. Data on other meats and poultry are not available. Recent consumption data and scientific evidence suggest that inclusion of red meat and processed meat at current intake levels is associated with positive health outcomes and improved diet quality [15]. Results from clinical trials reveal that lean red meat can be included in a cholesterol-lowering diet that reduces risk for cardiovascular disease [16,17].

3. MEAL PATTERNS WITH MEATS

Menu models using the 2010 USDA Food Pattern for a 2,000 calorie diet illustrate that menus incorporating commonly consumed foods and meals, including red and processed meats and foods eaten away from home, meet daily Myplate and energy goals, adhere to limits for saturated fat and sodium, and meet or exceed needs for key nutrients including nutrients of concern (Table 3). This modeling reinforces guidance emphasizing proper portion sizes and smart choices within food groups to build an overall eating pattern consistent with dietary guidance. Meats paired with other food groups, such as in sandwiches, salads, soups, and stews, can help increase intakes of whole grains, vegetables, and low-fat dairy. Research examining red meat consumption indicates that higher vegetable consumption is associated with eating meat [18].

Updated 2015 USDA Food Patterns at 12 calorie levels, using the most current food composition and food intake data available, meet nearly all nutritional goals for adequacy with the exception of potassium, vitamin D, vitamin E, and choline in almost all patterns [5]. Daily nutrient totals for the menu models indicate that most all nutrient goals were met within 75% of daily goals, illustrating that realistic menus consistent with USDA food patterns are attainable with a variety of popular food choices, including whole and convenience foods.

Table 3: Menu Models with Meat Based on Healthy U.S.-Style Pattern, 2,000

Meal	Menu: Day 1	Nutrient	Daily Total[1]	% Daily Value[2]
Breakfast	1 packet instant oatmeal made with: 1 cup fat-free milk 2 tbsp raisins 1 ½ tbsp brown sugar 1 cup orange juice	Calories Total fat (g) Saturated fat (g) Trans fat (g) Cholesterol (mg) Sodium (mg)	1,991 65 14 0 91 2,412	- 100% 68% - 30% 100%
Lunch	1 turkey hot dog 1 whole wheat hot dog bun ½ tsp yellow mustard 16 multigrain chips 1 cup fresh strawberry halves 12 fl oz sweetened tea	Potassium (mg) Protein (g) Carbohydrates (g) Total sugars (g) Dietary Fiber (g) Vitamin A (IU) Vitamin C (mg)	3,606 75 293 165 28 8,463 284	103% - 98% - 111% 169% 474%
Dinner	Chef's salad made with: 2 cups mixed baby greens ¼ medium red onion ½ medium tomato, chopped 2 oz turkey breast luncheon meat 1 ½ oz Swiss cheese, shredded 1 tbsp olive oil 1 tbsp balsamic vinegar 2 oz whole wheat dinner roll Water	Vitamin D (IU) Calcium (mg) Iron (mg) **Food Groups[1]**	250 1,500 17 **Amount**	63% 150% 96% **% of goal**
Snack	1 oz dry roasted, unsalted almonds 1 cup non-fat fruit yogurt with 2 tbsp granola	Grains Vegetables Fruits Milk Products Proteins	6.1 oz 2.6 cups 2.3 cups 3 cups 5.6 oz	102% 104% 115% 100% 102%

Meal	Menu: Day 2
Breakfast	1 whole wheat English Muffin 2 tbsp creamy peanut butter 1 cup cantaloupe pieces
Snack	1 container (5.3 oz) blueberry nonfat Greek yogurt 2 tbsp slivered almonds
Lunch	Spinach salad made with: 2 ½ cups baby spinach 2 slices bacon, broiled 1 large hard-boiled egg ¼ cup sliced mushrooms ¾ cup Mandarin oranges ¼ cup croutons 2 tbsp honey mustard dressing 1 cup fat-free milk
Dinner	Spaghetti and meatballs: 1 cup whole grain spaghetti, cooked ½ cup spaghetti sauce 2 oz meatballs (beef and pork) 2 tbsp grated Parmesan ½ cup green beans 4 fl oz red wine
Snack	4 graham cracker squares Chocolate milk made with: 1 ½ cups fat-free milk 2 tbsp chocolate syrup

Nutrient	Daily Total[1]	% Daily Value[2]
Calories	1,982	-
Total fat (g)	66	100%
Saturated fat (g)	14	70%
Trans fat (g)	0	-
Cholesterol (mg)	281	94%
Sodium (mg)	2,309	96%
Potassium (mg)	3,597	103%
Protein (g)	95	-
Carbohydrates (g)	243	81%
Total sugars (g)	127	-
Dietary Fiber (g)	27	108%
Vitamin A (IU)	14,751	295%
Vitamin C (mg)	142	237%
Vitamin D (IU)	343	86%
Calcium (mg)	1,538	150%
Iron (mg)	13	72%
Food Groups[1]	**Amount**	**% of goal**
Grains	6 oz	100%
Vegetables	2.7 cups	108%
Fruits	1.9 cups	95%
Milk Products	3.5 cups	116%
Proteins	5.7 oz	104%

Meal	Menu: Day 3	Nutrient	Daily Total[1]	% Daily Value[2]
	1 grapefruit	Calories	1,961	-
	Breakfast sandwich made with:	Total fat (g)	36	55%
Breakfast	1 large egg, fried	Saturated fat (g)	10	50%
	2 oz Canadian bacon	Trans fat (g)	0	-
	1 4-inch plain bagel	Cholesterol (mg)	327	109%
Snack	1 granola bar with fruit and nuts	Sodium (mg)	2,259	94%
	1 medium baked potato topped with:	Potassium (mg)	4,198	120%
	¼ cup reduced sodium black beans, canned	Protein (g)	102	-
		Carbohydrates (g)	316	105%
	1 oz part-skim mozzarella, shredded	Total sugars (g)	110	-
	Side salad made with:	Dietary Fiber (g)	27	108%
Lunch	1 cup mixed greens	Vitamin A (IU)	8,779	176%
	¼ cup cherry tomatoes	Vitamin C (mg)	190	317%
	1 tsp olive oil	Vitamin D (IU)	338	85%
	1 tsp balsamic vinegar	Calcium (mg)	1,319	110%
	4 low-sodium whole wheat crackers	Iron (mg)	14	78%
	1 ½ cup fat-free milk			
	Stir-fry made with:			
	3 oz skinless, boneless chicken breast			
	¼ cup broccoli			
	¼ cup sliced carrots			
Dinner	¼ cup sliced onion			
	¼ cup sliced mushrooms			
	1 tsp canola oil			
	1 tsp low-sodium soy sauce			
	1 cup cooked white rice			
	Water	**Food Groups[1]**	**Amount**	**% of goal**
		Grains	7 oz	116%
	Chocolate pudding made with 1 cup	Vegetables	3.5 cups	140%
Snack	fat-free milk	Fruits	2 cups	100%
	1 medium apple	Milk Products	3.1 cups	103%
		Proteins	6 oz	109%

[1] Data derived from menu analysis using The Food Processor, version 10.15.0 (ESHA Research, Salem, Oregon).
[2] Percent Daily Value (DV) are based on a 2,000 calorie diet.

4. NUTRIENT CONTRIBUTIONS OF MEATS

Meats make important nutrient contributions that if lowered in dietary patterns may result in unintended consequences of nutrient inadequacy, lower diet quality, and reduced compliance. Meats, including red meat, poultry, and processed meats, are nutrient-dense choices, serving as food sources of key nutrients, including high-quality protein, bioavailable iron and zinc, B vitamins, including thiamin, niacin, riboflavin, vitamin B6 and B12, and selenium and choline. For example, a 3-ounce serving of beef is considered an excellent source (20% or more of the Daily Value per serving) of protein, zinc, selenium, niacin, vitamin B6, and vitamin B12, and a good source (10-19% of the Daily Value per serving) of phosphorus, iron, riboflavin, and choline [19].

5. PROTEIN

Dietary intake data suggest that protein food intake is below recommendations for about 45% of Americans, with approximately 70% of adolescent girls and older women not meeting daily protein food recommendations [5]. Proposed healthy dietary patterns recommend protein intakes at 87-100 g/d for diets 1,800-2,200 calories/d, yet protein intakes for adult women average at 68 g/d, with adolescent girls at 63 g/d and older women (70+ years) consuming only 59 g/d [3]. The Acceptable Macronutrient Distribution Range (AMDR) for protein, established as the range associated with reduced risk for chronic disease while providing adequate intakes of essential nutrients, is 10-35% of energy intake for adults, or 50-175 g for a 2,000-calorie diet, with the lower end of the range set at approximately the Recommended Dietary Allowance (RDA), or 0.8-0.95 g/kg/d for ages 4 and older [20]. Most adults currently consume about 15% of their daily calories from protein [3].

Growing body of evidence suggests positive health outcomes associated with protein intakes at the higher end of the AMDR range or approximately 1.0-1.6 g/kg/d, and evenly distributing protein intake at meals throughout the day [21]. Dietary protein has been shown to be a key element of most effective weight loss regimens, with evidence for enhanced satiety, increased thermic effect of food, and preservation of lean body mass during weight loss [22-27]. Protein-induced satiety is believed to be an important component of weight loss and maintenance, with evidence for higher protein meals enhancing satiety measures for several hours after a meal [24,25,28,29]. Higher protein weight loss diets (1.2-1.6 g/kg/d) resulted in greater losses in body weight and body fat, preservation of lean body mass, and reductions in triglycerides, blood pressure, and waist circumference [25,30,31]. During weight maintenance, higher protein diets favor regain of fatfree mass over fat mass, and reduce the efficiency of weight gain [22].

Food sources of protein differ in their protein content, and in the quality of protein provided, as measured by the provision of EAAs. Animal proteins are known to provide a higher quality and greater quantity of protein compared to plant proteins, and often provide a more energy efficient source of protein. For example, to obtain the same amount of protein found in lean meat from most plant proteins, such as beans, nuts and grains, calories may be 1½ to 2 times greater (Table 4). According to an analysis of food sources of energy and nutrients among US adults, poultry and beef are among the three highest ranked food sources of protein; beef was also within the top three food sources of total fat and saturated fat [32]. Plant sources of protein included yeast bread and rolls ranked as number 5 and nuts and seeds at number 12 [32].

Dietary guidance emphasizing nutrient density supports a similar approach to categorize a protein source's ability to provide EAAs in an efficient manner specific to its energy content [33]. This proposed concept of EAA density underscores the critical role dietary protein has in preventing, slowing the progression of, or managing chronic disease. Consuming high-quality protein with optimum ratios of essential amino acids at multiple meals each day has been shown to support metabolic health [29,34]. Essential amino acids act as metabolic signals, influencing protein synthesis, inflammation responses, and satiety. Leucine, an EAA found in higher amounts in animal proteins, has been found to stimulate muscle protein synthesis and may improve satiety and insulin sensitivity [35-39]. In older adults, protein intake may reduce the risk of sarcopenia, or the gradual loss of muscle mass and function. Evidence indicates that consuming a moderate amount of high-quality protein (1.0-1.5 g/kg/d) evenly distributed at each eating occasion (i.e., 20-30 g/meal), combined with physical activity, can be an effective strategy for adults to maximize lean body mass before its decline, and help slow the progression of or prevent sarcopenia in older adults [30,40-42].

Table 4: Energy Comparison of 25-gram Protein Sources.

Food Source and Amount	Calories	Protein (g)
Lean sirloin, 3 oz cooked	160	26
Skinless chicken breast, 3 oz cooked	137	26
Pork loin chop, 3 oz cooked	148	25
Red beans, 1.6 cups cooked	366	25
Firm tofu, 10 oz	235	25

Calories required to obtain approximately 25 grams protein (cooked or ready-to-eat portions)
Source: United States Department of Agriculture Agricultural Research Service National Nutrient Database for Standard Reference Release 27

6. MICRONUTRIENTS

In addition to high quality protein, meat and poultry are important sources of micronutrients, including iron, zinc, selenium, and vitamins B12, B6, thiamin, riboflavin, niacin, and potassium. Iron is recognized as being under consumed by adolescent and premenopausal women, including women who are pregnant. The 2015 DGAC Report recognizes that red meats are an excellent source of heme iron, a highly bioavailable form of iron. Plant sources of iron may contain phytates, tannins, and other compounds that reduce absorption of non-heme iron. Consuming meat, poultry, and vitamin C with a plant-source of iron, and soaking beans and grains in water can help to increase absorption of iron from plant sources. Research suggests that at least 50 g or about 2 oz of meat or poultry may improve iron absorption from a meal by approximately 45% [43].

Up to 16% of US adults and more than 20% of adults over 60 years old are marginally depleted in vitamin B12, with inadequate intake due to low consumption of animal-source foods as a primary cause of low serum vitamin B12 [44]. Animal foods are an excellent source of vitamin B12, an essential vitamin not naturally available in plant protein sources. Deficiency of this vitamin increases with age, with the added complication of gastric atrophy with aging, which reduces absorption of vitamin B12 in older adults. The decline in protein food intake observed with age may exacerbate the increased risk of vitamin B12 seen in older adults. Of adults 71 years and older, only about 30% of women and 50% of men meet the daily recommendation for Protein Foods [5]. An analysis of food sources of energy and nutrients among US. adults indicates that beef is the highest ranked food source of vitamin B12, while beef and poultry rank among the top 5 sources of vitamin B6, zinc, and phosphorus [32]. Beef, poultry, pork, and processed meats are ranked at numbers 6, 8, 12, and 14 respectively, as top food sources of potassium, a nutrient of public health concern recognized by the 2015 DGAC [5,32].

In regards to sodium, meat and poultry processing often uses salt for flavoring and to enhance food safety, making these options higher in sodium compared to fresh meats. According to the North American Meat Institute, meat and poultry processors are actively engaged in efforts to reduce sodium and currently offer a variety of choices with different sodium contents, including 'reduced sodium' with a 75% reduction from the original formulation, and 'low sodium' with 140 mg of sodium or less per serving [45].

7. POSITIVE AND REALISTIC GUIDANCE ON MEAT INTAKE

Consumer inertia toward healthier eating, or the tendency to do nothing or remain unchanged, is a significant challenge in moving consumer behaviors closer to achieving dietary guidance goals [46]. Changes in nutrition guidance and contradictory nutrition messages have led to consumer confusion, with negative advice to avoid or limit foods likely contributing to a backlash against nutrition recommendations [47,48]. For food guidance to have the intended effect, it must be positive and realistic, without an expectation that consumers must surrender the foods they love. The effectiveness of dietary guidance is contingent on advice that is also consistent, easy-to-understand, and actionable, personalized and fits within consumers' time, money, and skill constraints [49]. Related to meat intake, guidance that reinforces consumers' preferences and provides targeted and tailored advice for subgroups that may be under or over consuming meat or selecting less healthful options is critical to help consumers achieve a balanced lifestyle with better health outcomes.

Although consumers want to maintain control over their food choices, they are receptive to messages that empower, encourage, reassure, and offer advice that fits within their everyday lives [49]. Traditional "one size fits all" nutrition advice may have contributed to the lack of motivation and apathy around changing eating behaviors, while targeted and tailored nutrition messages are more likely to be perceived as personally relevant and achievable [50,51]. Behavioral research also suggests that positive and practical nutrition education is a more effective approach to behavior change than penalizing with taxes and restrictions. A 2013 nationally representative survey of 1,011 US adults found that 83% of those surveyed were in favor of the government providing nutritional guidelines and information about how to make healthy choices, yet there is little support for policies that would constrain or penalize consumer choices [52,53]. Seventy-four percent strongly or somewhat opposed limiting the types or amounts of food and drinks that people can buy, and 59 percent opposed placing a tax on the sale of unhealthy foods and drinks [53].

Positive messages that show consumers how to choose and incorporate lean meats, including red and processed meats, are more likely to inspire positive behavior change. Providing consumers with alternative options, such as lean cuts of meat and products with reduced fat or sodium that can be tailored to personal preferences may improve the likelihood of long-term success in maintaining a healthy dietary pattern. Guidance supported with education about choosing lean and lower sodium meats, both fresh and processed, can help consumers balance their diets and lifestyle, practice portion control, and meet their energy and nutrient needs.

8. CONCLUSION

While it is important to consume protein from a variety of animal-and plant-based sources, animal foods generally provide more protein by weight than plant foods and supply higher quality protein, which are concepts warranting greater attention in dietary guidance. Furthermore, fewer calories are needed to achieve adequate protein intake from animal-based compared to plant-based foods.

Consumption data indicates that Americans are eating meat in moderation the only MyPlate food group being consumed within the recommended daily amounts in USDA food patterns. Protein is not considered a shortfall nutrient, with dietary intake data indicating that Americans are meeting the RDA for protein, which falls at the low end of the AMDR. However, significant and emerging evidence supports the role of high-quality protein intakes at the mid-to-high end of the AMDR in weight management, metabolic function, healthy aging, and nutrient adequacy, which translates to intakes of protein of up to twice the RDA from both animal and plant sources, and balancing intake throughout the day to optimize these benefits [33].

The term "red meat" is ambiguous, both in research and in dietary guidance, and may or may not include "processed meat," which also lacks a universal definition. In research, the lack of a standard definition for red meat means that measurement of consumption is subject to either individual study participant interpretation of what constitutes red meat or differing definitions provided by researchers, both of which may account for the conflicting research results on the effects of red and processed meat on health. The lack of a uniform strategy for characterizing and analyzing red and processed meat intake limits the possibility of comparing study findings and drawing clear and consistent conclusions within and across studies of the associations between meat and health outcomes. The 2015 DGAC acknowledged that further evaluation of studies and the definition of meats are needed to provide evidence of the differences between certain foods and relationships to health outcomes [5]. Research derived from a standardized definition of red meat that differentiates between various types (i.e., lean, processed) and is consistently used across studies is more likely to produce results that can be used to inform scientifically accurate messages about red and processed meat in education, outreach, and nutrition policy-making efforts.

A recommendation to reduce meat intake may be detrimental, with the potential consequence of lowering protein and nutrient intakes, and reducing satisfaction and compliance with healthpromoting diets due to physiological and/or psychological parameters associated with limiting protein- and nutrient-rich meats. Advice to successfully change eating habits must reflect how Americans eat, and promote a positive approach for how to include desirable foods rather than constraining consumer choice. While there is broad support for government policies that would facilitate a healthier lifestyle, the majority of consumers draw the line at policies that limit choice [53]. As noted by the 2015 DGAC, a range of diverse foods can be included to achieve healthful eating patterns, consistent with personal, cultural, and religious preferences [54]. This and other advice recognizing that there are many ways and foods that can be included in a healthful diet aligns with consumer research supporting positive nutrition messages that target and tailor nutrition advice to specific audiences.

Menu models demonstrate that a variety of red and processed meats can be incorporated into a healthful eating pattern that meets energy and nutrient recommendations from the 2010 Dietary Guidelines for Americans and MyPlate food groups. Recognizing that many Americans enjoy red and processed meats, recommendations that limit or restrict this protein-and nutrient-rich food may threaten the acceptance and implementation of future Dietary Guidelines. High quality protein, including lean red meat and reduced-fat or low-sodium processed meat, has a critical role in human health and nutrition, and continued promotion of lean meats has the potential to motivate behavior change and increase success in achieving healthful dietary patterns.

ACKNOWLEDGMENT

Funding provided by North American Meat Institute. Ms Hornick has declared that she is a member of FoodMinds' expert network. FoodMinds is a food and nutrition affairs company that represents a range of food, nutrition, and wellness organizations, including the North American Meat Institute. Ms Cochran received an honorarium for her contributions. Both authors contributed to this manuscript and approved the final version of this paper.

REFERENCES

1. Zanovec M, O'Neil CE, Keast DR, Fulgoni VL, Nicklas TA (2010) Lean beef contributes significant amounts of key nutrients to the diets of US adults: National Health and Nutrition Examination Survey 1999-2004. Nutr Res 30: 375-381.

2. Consumption Habits Survey (2012) Gallup Poll.

3. What we Eat in America, NHANES 2011-2012, individuals 2 years and over (excluding breast-fed children), day 1 dietary intake data, weighted. Food Patterns Equivalents Database (FPED) 2011-2012.

4. US Department of Agriculture; US Department of Health and Human Services (2010) Dietary Guidelines for Americans (7thedn) US Government Printing Office: Washington, DC.

5. US Department of Agriculture (2015) Center for Nutrition Policy and Promotion. Report of the Dietary Guidelines Advisory Committee on the Dietary Guidelines for Americans.

6. Davis C, Saltos E (1999) Dietary Recommendations and How They Have Changed Over Time. Ed. Frazao E. America's Eating Habits: Changes and Consequences. Agriculture Information Bulletin No. AIB-750.

7. National Cattlemen's Beef Association (2014) Lean Matters: Chronicling Beef's Change from Gate to Plate.

8. Food Marketing Institute, North American Meat Institute (2015) The Power of Meat: An In-depth Look at Meat Through the Shoppers' Eyes.

9. Bauer J, Biolo G, Cederholm T, Cesari M, Cruz Jentoft AJ, et al. (2013) Evidence-based recommendations for optimal dietary protein intake in older people: a position paper from the PROT-AGE Study Group. J Am Med Dir Assoc 14: 542-559.

10. Morrison RM, Buzby JC, Wells HF (2010) Guess Who's Turning 100? Tracking a Century of American Eating. Amber Waves.

11. US Department of Agriculture Economic Research Service, Food Availability (Per Capita) Data System: Food availability and dietary trends.

12. US Department of Agriculture, Economic Research Service, Food Availability (Per Capita) Data System: Loss-Adjusted Food Availability.

13. US Department of Agriculture, Center for Nutrition Policy and Promotion. Nutrient Content of the US Food Supply, 1909-2010.

14. Barraj L, Smith K, Scrafford C (2010) Nutritional contributions of pork to the diets of the US population. Abstract 560.3. FASEB J.

15. Fulgoni V, Agerwal S, Berg E Association of lunch meat consumption with diet quality in children and adults: NHANES 2007-2010.

16. Hunninghake DB, Maki KC, Kwiterovich PO, Jr. Davidson MH, Dicklin MR, et al. (2000) Incorporation of lean red meat into a National Cholesterol Education Program Step I diet: a long-term, randomized clinical trial in free-living persons with hypercholesterolemia. J Am Coll Nutr 19: 351-360.

17. Roussell MA, Hill AM, Gaugler TL, West SG, Heuvel JP, et al. (2012) Beef in an Optimal Lean Diet study: effects on lipids, lipoproteins, and apolipoproteins. Am J Clin Nutr 95: 9-16.

18. Kappeler R, Eichholzer M, Rohrmann S (2013) Meat consumption and diet quality and mortality in NHANES III. Eur J Clin Nutr 67: 598-606.

19. US Department of Agriculture, Agricultural Research Service (2014) USDA National Nutrient Database for Standard Reference, Release 27. Nutrient Data Laboratory Home Page.

20. Institute of Medicine, Board, FAN (2005) Dietary Reference Intakes for Energy, Carbohydrates, Fiber, Fat, Fatty Acids, Cholesterol, Protein, and Amino Acids. Washingon DC: National Academies Press.

21. Rodriguez NR (2015) Introduction to Protein Summit 2.0: continued exploration of the impact of high-quality protein on optimal health. Am J Clin Nutr 101: 1317S-1319S

22. Westerterp P, Nieuwenhuizen MA, Tome D, Soenen S, Westerterp KR (2009) Dietary protein, weight loss, and weight maintenance. Annu Rev Nutr 29: 21-41.

23. Paddon-Jones D Westman E, Mattes RD, Wolfe RR, Astrup A, et al. (2008) Protein, weight management, and satiety. Am J Clin Nutr 87: 1558S-1561S.

24. Leidy HJ, Clifton PM, Astrup A, Wycherley TP, Westerterp-Plantenga MS, et al. (2015) The role of protein in weight loss and maintenance. Am J Clin Nutr 101: 1320S-1329S

25. Leidy HJ, Carnell NS, Mattes RD, Campbell WW (2007) Higher protein intake preserves lean mass and satiety with weight loss in pre-obese and obese women. Obesity (Silver Spring) 15: 421-429.

26. Astrup A, Raben A, Geiker N (2014) The role of higher protein diets in weight control and obesity-related comorbidities. Int J Obes (Lond) 39: 721-726.

27. Veldhorst M, Smeets A, Soenen S, Hochstenbach-Waelen A, Hursel R, et al. (2008) Protein-induced satiety: effects and mechanisms of different proteins. Physiol Behav 94: 300-307.

28. Leidy HJ, Ortinau LC, Douglas SM, Hoertel HA (2013) Beneficial effects of a higher-protein breakfast on the appetitive, hormonal, and neural signals controlling energy intake regulation in overweight/obese, "breakfast-skipping," late-adolescent girls. Am J Clin Nutr 97: 677-688.

29. Leidy HJ, Tang M, Armstrong CL, Martin CB, Campbell WW (2011) The effects of consuming frequent, higher protein meals on appetite and satiety during weight loss in overweight/obese men. Obesity (Silver Spring) 19; 818-824.

30. Paddon-Jones D, Leidy H (2014) Dietary protein and muscle in older persons. Curr Opin Clin Nutr Metab Care 17: 5-11.

31. Bosse JD, Dixon BM (2012) Dietary protein in weight management: a review proposing protein spread and change theories. Nutr Metab (Lond) 9: 81.

32. O'Neil CE, Keast DR, Fulgoni VL, Nicklas TA (2012) Food sources of energy and nutrients among adults in the US: NHANES 2003-2006. Nutrients 4: 2097-2120.

33. Rodriguez NR, Miller SL (2015) Effective translation of current dietary guidance: understanding and communicating the concepts of minimal and optimal levels of dietary protein. The American Journal of Clinical Nutrition.

34. Mamerow MM, Mettler JA, English KL, Casperson SL, Arentson-Lantz E, (2014) Dietary protein distribution positively influences 24-h muscle protein synthesis in healthy adults. J Nutr 144: 876-880.

35. Churchward-Venne TA, Murphy CH, Longland TM, Phillips SM, (2013) Role of protein and amino acids in promoting lean mass accretion with resistance exercise and attenuating lean mass loss during energy deficit in humans. Amino Acids 45: 231-240.

36. Norton LE, Layman DK (2006) Leucine regulates translation initiation of protein synthesis in skeletal muscle after exercise. J Nutr 136: 533S-537S.

37. Layman DK (2003) The role of leucine in weight loss diets and glucose homeostasis. J Nutr 133: 261S-267S.

38. Burke LM, Winter JA, Cameron-Smith D, Enslen M, Farnfield M, (2012) Effect of intake of different dietary protein sources on plasma amino Acid profiles at rest and after exercise. Int J Sport Nutr Exerc Metab 22: 452-462.

39. Esteves de Oliveira FC, Pinheiro Volp AC, Alfenas RC (2011) Impact of different protein sources in the glycemic and insulinemic responses. Nutr

40. Breen L, Phillips SM (2011) Skeletal muscle protein metabolism in the elderly: Interventions to counteract the 'anabolic resistance' of ageing. Nutr Metab (Lond) 8: 68.

41. Beasley JM, Shikany JM, Thomson CA (2013) The role of dietary protein intake in the prevention of sarcopenia of aging. Nutr Clin Pract 28: 684-690.

42. Paddon-Jones D, Campbell WW, Jacques PF, Kritchevsky SB, Moore LL, et al. (2015) Protein and healthy aging. Am J Clin Nutr.

43. Lim KH, Riddell LJ, Nowson CA, Booth AO, Szymlek-Gay EA (2013) Iron and zinc nutrition in the economically-developed world: a review. Nutrients 5: 3184-3211.

44. Allen LH (2009) How common is vitamin B-12 deficiency? Am J Clin Nutr 89: 693S-696S.

45. American Meat Institute. Processed Meats: Convenience, Nutrition, Taste.

46. Webb D, Byrd-Bredbenner C (2015) Overcoming consumer inertia to dietary guidance. Adv Nutr. In press.

47. Nagler RH (2014) Adverse outcomes associated with media exposure to contradictory nutrition messages. J Health Commun 19: 24-40.

48. Watts ML, Hager MH, Toner CD, Weber JA (2011) The art of translating nutritional science into dietary guidance: history and evolution of the Dietary Guidelines for Americans. Nutr Rev 69: 404-412.

49. Rowe S, Alexander N, Almeida NG, Black R, Burns R, et al. (2011) Translating the Dietary Guidelines for Americans 2010 to bring about real behavior change. J Am Diet Assoc 111: 28-39.

50. Bier D, Derelian D, German J, Katz D, Pate R, et al. (2008) Improving compliance with dietary recommendations. Nutr Today 43: 180-187.

51. Hornick BA, Childs NM, Smith EM, Kapsak WR, Dooher C, et al. (2013) Is it time to rethink nutrition communications? A 5-year retrospective of Americans' attitudes toward food, nutrition, and health. J Acad Nutr Diet 113: 14-23.

52. Gollust SE, Barry CL, Niederdeppe J (2014) Americans' opinions about policies to reduce consumption of sugar-sweetened beverages. Prev Med 63: 52-57.

53. The Associated Press- NORC (2013) Center for Public Affairs Research. University of Chicago. Obesity in the United States: Public Perceptions.

54. Post R, Budak C, Canavan J, (2007) A Guide to Federal Food Labeling Requirements for Meat, Poultry, and Egg Products. USDA 2007.

9

UTILIZATION OF YOGHURT AND SUCRALOSE TO PRODUCE LOW-CALORIE CAKES

Hussein AMS*, Hegazy NA, Kamil MM and Ola SSM

Food Technology Department, National Research Centre, Dokki, PO box 12311, Giza, Egypt

ABSTRACT

In this study we had replaced sucrose with sucralose in the manufacture of sponge cake and yoghurt cake. These cakes were chosen because their ingredients include little fat so they were of low calories. Such functional cakes were evaluated chemically, physically and sensorial. The results suggest that, sucrose substitution with sucralose in cakes increased the cake volume and softened the texture (as shown by lower values of hardness, chewiness and gumminess). Water activity of the yoghurt control cake prepared with 100% sucrose with a value of 0.91 was higher than the yogurt cake containing sucralose ($p<0.05$). The mean water activity of sponge cakes prepared with sucrose was significantly higher than the sponge cakes containing sucralose ($p<0.05$). The obtained low calorie yoghurt cake characterized with its low calorie and food energy (103.22 and 431.3 per100 g) than control yoghurt cake (268 and 1119 per 100 g). Also, low calorie sponge cake had low calorie and food energy (98.0 and 409.4 per100 g) compared to control sponge cake (276.9 and 1156.9, respectively). Sensory scores of studied cake samples indicated that, texture and flavor of all samples were not affected significantly in case of replacing sucrose with sucralose in yogurt or sponge cake. A significant difference in cells and crumb color was observed in all cake samples when compared with control cake samples.

KEYWORDS

Functional cake; Sucralose; Low calorie cake; Sponge cake; Yoghurt cake; Texture profile; Calorie values

1. INTRODUCTION

Cake is one of baked products, which generally contains fat, sugar, wheat flour, egg and milk. It is composed of 20-50% fat and 10-30% sugar depending on types of cakes. Due to its high fat and high calorie content, overconsumption may contribute to high risk of health problems such as coronary diseases, high blood pressure and cholesterol, diabetes, obesity and some cancers [1,2].

Low calorie foods available to consumer shelves on the market are products prepared with low energy sweeteners. These products are very popular among weight and health conscious consumers [3]. Generally, fat reduction is the primary target prior to the replacement of sucrose in the production of low-calorie foods. The replacement of sucrose by using these sugar substitutes alone (xylitol, sorbitol, acesulfame-K, aspartame, saccharin) or their combinations at levels of above 25% resulted in a decrease in the quality and acceptability of cakes [4,5]. Consequently, it is important to find an alternative sugar substitute to traditional sugars in order to improve the quality of cakes.

Sucralose was discovered by British researchers in 1976. It is a chlorinated disaccharide [6]. Sucralose (1,6-dichloro-, 1,6-dideoxy-β-Ddructofuranosyl-4-chloro-deoxy-α-D-galactopyranoside) is synthesized from sucrose by selectively replacing 3 hydroxyl groups with chlorine atoms [7,8]. It is a zero-calorie sweetener that is 600 times sweeter than sucrose [9] but can range from 400-800 times sweeter, depending on desired sweetness and product formulation [10]. Sucralose tastes like sugar, has no unpleasant or bitter aftertaste [11], and is non-calorie, non-carcinogenic, and proven safe for human consumption [12]. Sucralose contains no calorie because it is not hydrolyzed in the intestinal lumen, not converted to energy in the body. It is poorly absorbed by experimental animals and is excreted largely unchanged in the feces [10]. Sucralose is highly soluble in water and alcohol [11], chemically compatible with all common food ingredients, and stable in a wide range of pH levels and thermal processes [11]. Thus, it could be a suitable sweetener for bakery products [13]. Although sucralose provides the sweetness and crystallization properties of sucrose, it cannot mimic the structural contribution of sucrose to baked products [9]. It cannot create bulk properties in the cake batter as sucrose does. As a result, sucralose must be combined with other substances to match sucrose's bulking properties. A few previous studies reported that replacing sucrose with sucralose up to 50% has produced acceptable products: chiffon cake formulated with a mixture of sucralose and a type of indigestible dextrin in chiffon cakes [7], reduced-fat chiffon cake containing 50% erythritol-sucralose [14], and muffins using sucralose and polydextrose [9]. Sweeteners were successful in making the texture of cakes softer and the quality of the texture of the cake did not change for a long time after baking [15]. Therefore, this study aimed to produce and evaluate low calorie yoghurt and sponge cakes free from shortening and sucrose that could be used as a functional cake for diabetes and obesity.

2. MATERIALS AND METHODS

2.1 Materials

Soft wheat flour (Gemiza 9), 72% extraction was obtained from the Cairo south Company of milling (EL-Haram Milling). Yoghurt, sucralose, ground sugar, vanilla, eggs and baking powder were obtained from the local market, Cairo, Egypt.

2.2 Methods

Yoghurt Cake preparation: Yogurt, sugar, and eggs were whipped at medium speed using a kitchen mixer (Moulinex, Model HM1010, Beijing china) for 7 minutes then vanilla was added. In a medium bowl, whisk together flour, baking powder, and salt. In a separate bowl, whisk together yoghurt, Sugar, eggs, and vanilla were mixed well with wheat flour and passed through a stainless steel screen and then added to the mixture. Sample of cake was placed into rectangular metallic pans (80 mm width, 175 mm length and 50 mm depth) and baked at 180°C in an electric oven for 35 minutes. Cakes were removed from the pans and left at room temperature for one hour. The cakes were then sealed in polyethylene bags to be prevented from becoming dry. For the manufacture of cakes for diabetics follow the same steps in the normal cake with sugar replacement with sucralose.

Preparation and evaluation of sponge cakes: Sponge cakes were prepared according to Bennion and Pamford [16] with some modifications as follows: Flour (100 gm) and baking powder (3 gm) were mixed together; whole fresh eggs (125 gm), and sugar (100 gm) were whipped for 6 min. using a mixer at high speed, then vanilla was added. Flour mixture was added gradually on mixture and beaten for 3 min. using the mixer at low speed. Cake were poured in baking pan (80 mm width, 175 mm length and 50 mm depth), then placed in a preheated oven and baked at 180°C for 35 min. Cakes were allowed to cool for 30 min. in the pans at room temperature. Low calorie cakes were prepared using the same steps in the common cake (control) with replacing sucrose with sucralose

Physico-chemical analysis: All cakes and wheat flour (72%) were homogenized to prepare the samples for analysis. Moisture, protein, lipid, ash and carbohydrate were determined according to AOAC [17]. Water activity of the cakes was determined by using an Aqua Lab device (Model CX2, Decagon Device, and Pullman, WA).

Specific volume was determined according to AACC [17], where rapeseed displacement was used to measure cake volume. The cake volume was divided by cake weight to express the specific volume of the cake.

Color parameters of all samples were determined using a spectrocolorimeter (Tristimulus Colour Machine) with the CIE lab colour scale (Hunter, Lab Scan XE-Reston VA, USA) in the reflection mode. The instrument was standardized with white tile of Hunter Lab Colour Standard (LX No.16379). The color parameters are based on the Hunter L*, a* and b* coordinates. The L scale ranges from 0 black to 100 white; the a* scale extends from a* negative value (green hue) to a positive value (red hue) and the b* scale ranges from negative blue to positive yellow.

Texture profile analysis of cakes Hardness and resilience analysis was conducted using Brookfield Texture Analyzer. The analyzer was set to perform single cycle measurements which are used for the determination of the first bite force of cakes. The measurement speed of 2 mm/s and a distance of 5 mm were applied. A force-time diagram was taken for each test. The force-time plots were analyzed for peak breaking force (g) and time (s) to reach the peak. Textural attributes were measured in three independent samples and the presented values are mean values.

Differential scanning calorimetry (DSC) measurements were performed with Shimadzu DSC-50 in the four blends and in the corresponding biscuits of the control and the TNFcontaining biscuits. The samples were heated to 200°C at a 10°C/min heating rate. The onset temperature (To), peak temperature (Tp), end set temperature (Te) and (ΔH) of the enthalpy was calculated from the DSC thermo grams. The enthalpy was expressed in J/g of baked cakes.

2.3 Determination of energy

Calculation from the following equation: Calculated Energy=(Protein + carbohydrate) × 4 + fat × 9.

Calorimeter device: Differential scanning calorimeters, isothermal micro calorimeters, titration calorimeters and accelerated rate calorimeters are among the most common types.

Sensory evaluation: Cake samples were evaluated for the following parameters according to the method described in AACC [17]: cells (30), grain (16), texture (34), crumb color (10) and flavor (10).

Freshness of Cakes: Cakes freshness was tested after wrapping with polyethylene bags and stored at room temperature for 0, 1 and 3 days. Alkaline Water Retention Capacity test (AWRC) was carried out according to the method of Yamazaki [18]; and modification by Kitterman & Rubenthaler [19].

Statistical analysis: The obtained results were evaluated statistically using analysis of variance as reported by McClave & Benson [20].

3. RESULTS AND DISCUSSION

3.1 Chemical composition and nutritional evaluation

Data presented in Table 1 compared between gross chemical compositions and energy value of common yoghurt cake, sponge cake (control 1 and 2) and their low calorie cakes products (S1 and S2). The obtained results showed that moisture content of control samples in yoghurt (25.7%) or sponge cake (28%) was higher than low calorie cakes products of S1 (21%) or S2 (22%). Consequently, protein and ash of low calorie cakes products were increased. Also, replacing sucrose with sucralose in cakes led to decrease carbohydrates in cake of S1and S2 to 13.5 and 12%, while control samples of yoghurt and sponge cake reached to 54.3 and 56%, respectively.

Fat content of the cake was minimized in order to lower its calorie value, since it is well documented that obese and diabetic people have a tendency towards hypercholesterolemia. Therefore, yoghurt and sponge cakes samples were prepared without using shortening then sucrose was replaced with sucralose to offer cake suitable for diabetics. Table 2 showed the calculated (K. Cal/100 g) and determined energy (kJ/100 g) of studied cakes samples. The obtained results indicated that, replacing sucrose with sucralose in yoghurt cake (S1) decreased its food energy to 408.8 K.Cal/100 g, while control yoghurt cake reached to 1050 K.Cal/100 g. Also, replacing sucrose with sucralose in sponge cake (S2) decreased its food energy to 390.4 K.Cal/100 g compared to control sponge cake which reached to 1082.4 K.Cal/100 g. Consequently, low calorie sponge cakes and low calorie yoghurt cakes could be recommended to obese and diabetic people.

3.2 Color quality of cakes

Color is one of the most important sensory attribute that affect directly the consumer preference of any product. Special attention should be given to bakery products to attract the consumer attention. The color parameters of cake samples were evaluated using a Hunter laboratory colorimeter. Table 3 showed that low-calorie cake products (S1 & S2) were darker than control cakes (1 & 2), where lightness (L*), redness (a*) and yellowness (b*) of low-calorie cakes (S1 & S2) decreased compared to the control cakes (1 & 2). Such findings are inagreement with Kim et al. [21], Kordonowy &Young [22] and Ramy [23].

3.3 Textural properties of cakes

Texture profile analysis (TPA) is a very useful technique for investigating the physical properties of food products. In the present study, the TPA parameters of control samples in yoghurt, sponge cakes (1 & 2) and their cakes of replacing sucrose with sucralose (S1 & S2) were evaluated using texture analyzer by double compression tests, and presented in Table 4. Hardness was defined as the maximum force of the first compression of the product at the point of 50% compression (12.5 mm) of the original sample height. The hardness values of baked cakes increased significantly with the replacement of sucrose with sucralose. However, cohesiveness determined from the area of work during the second compression divided by the area of work during the first compression [24]. There was no significant difference between cohesiveness of cake samples with or without sucralose, where they ranged between 0.52-0.56. Springiness was defined as the distance to which the sample recovered in height during the time that elapsed between the end of the first compression cycle and the start of the second compression cycle.

Table 1. Chemical composition of yoghurt cake (control 1), sponge cake (control 2) and their cakes after replacing sucrose by sucralose, Mean values ± Standard Deviation, S1=Yoghurt cakes of replaced sucrose by sucralose, S2=Sponge cakes of replaced sucrose by sucralose, *Calculated Energy= (Protein + carbohydrate) x 4 + fat x 9, **Determined Energy=using calorimeter apparatus.

Samples	Moisture (%)	Protein (%)	Fat (%)	Ash (%)	Carbo-hydrate (%)	Energy	
						Calculated*K.Cal/100g	Determined**kJ/100g cake
Yoghurt cakes							
Control (1)	25.7 ± 0.15	6.8 ± 0.11	2.0 ± 0.05	1.4 ± 0.01	54.3 ± 0.32	262.4	1050 ± 0.59
S1	21.0 ± 0.19	7.2 ± 0.13	2.1 ± 0.02	1.8 ± 0.01	13.58 ± 0.12	102.02	408.8 ± 0.62
Sponge cakes							
Control (2)	28.0 ± 0.14	7.6 ± 0.05	1.8 ± 0.01	1.5 ± 0.02	56 ± 0.34	270.6	1082.4 ± 0.76
S2	22.0 ± 0.12	7.9 ± 0.08	2.0 ± 0.02	1.7 ± 0.01	12.0 ± 0.09	97.6	390.4 ± 0.52

Table 2. Formulations of yoghurt cake, sponge cake samples and their cakes after replacing sucrose by sucralose. S1=Yoghurt cakes of replaced sucrose by sucralose, S2=Sponge cakes of replaced sucrose by sucralose.

Ingredients	Yoghurt cake control (1)	Yoghurt cake (S1)	Sponge cake control (2)	Sponge cake (S2)
Wheat flour (gm)	100	100	100/ 29%	100 / 38.51%
Sucrose (gm)	75	- - -	100/ 29%	- - -
Succharalose (gm)	- - -	12.5	- - -	12.5
Yoghurt (gm)	225	225	- - -	- - -
skim milk (gm)	- - -	- - -	20	20
Whole Egg (gm)	56	56	125	125
Baking powder (gm)	3	3	0.5	0.5
Vanilla (gm)	0.65	0.65	0.65	0.65
Salt (gm)	1	1	1	1

Table 3. Cake color parameters of yoghurt cake (control 1), sponge cake (control 2) and their cakes after replacing sucrose with sucralose, S1=Yoghurt cakes of replaced sucrose by sucralose, S2=Sponge cakes of replaced sucrose by sucralose, *ΔE=(L2 + a2 + b2)1/2.

Samples	crust				Crumb			
	L	a	b	ΔE*	L	a	B	ΔE*
Yoghurt cakes								
Control (1)	55.29	16.94	42.16	71.56	67.27	5.56	37.14	77.04
S1	53.15	18.19	44.56	71.70	67.43	7.36	35.8	76.70
Sponge cakes								
Control (2)	70.13	5.31	40.87	81.34	73.04	1.72	40.35	83.46
S2	57.7	17.15	42.36	73.61	70.92	5.1	38.21	80.72

Springiness of testing cakes not affected significantly by replacing sucrose with sucralose in yoghurt cakes or sponge cakes. Gumminess was calculated by the product of (that is by multiplying) hardness and cohesiveness, whereas chewiness, defined as the energy required to masticate solid food to a state of readiness for swallowing [25] was obtained from the product of hardness, cohesiveness and springiness. Chewiness and gumminess values in cakes with and without sucralose exhibited a similar trend with the hardness values, as shown in Table 4. Significantly lower chewiness and gumminess were found in the cake samples containing sucralose. The obtained results suggest that partial sucralose in cakes increased the cake volume and softened the texture (as shown by lower values of hardness, chewiness and gumminess).

3.4 Differential scanning calorimetry

Differential scanning calorimetry of of yoghurt cake (control 1), sponge cake (control 2) and their cakes after replacing sucrose with sucralose (S1 & S2) are clearly shown in Table 5. Onset temperature (To), peak temperature (Tp) and enthalpy change ΔH (J/g) were automatically calculated using Universal Analysis software published by TA Instruments, New Castle, Delaware. As the amount of sucrose in formulation was decreased, both the onset and peak temperatures decreased. Control cakes had the highest onset and peak temperature; 72.64°C and 74.44°C, respectively. The treatment with 100% sucrose replacement showed the lowest onset (69.65°C) and peak (69.81°C) temperatures. The same results of onset and peak temperature in the sponge cake were noticed. These results correspond with differential scanning calorimetry testing conducted by Lim [26] on wheat starch, sucrose and water interactions. The researchers reported that the increase of onset temperature as sucrose concentrations were increased [26].

Table 4. Texture profile analysis parameters of yoghurt cake (control 1), sponge cake (control 2) and their cakes after replacing sucrose with sucralose (S1 & S2), S1=Yoghurt cakes of replaced sucrose by sucralose, S2=Sponge cakes of replaced sucrose by sucralose.

Samples	Hardness(N)	Cohesiveness	Springiness(mm)	Gumminess(N)	Chewiness(Nm)
Yoghurt cakes:					
Control (1)	120	288	2.4	390	320
S1	130	285	2.28	400	380
Sponge cakes:					
Control (2)	125	330	2.64	450	410
S2	140	332	2.60	410	380

Table 5. Thermal properties of yoghurt and sponge cakes and their low-calories cakes (S1&S2), Mean values ± Standard Deviation, S1=Yoghurt cakes of replaced sucrose by sucralose, S2=Sponge cakes of replaced sucrose by sucralose, To=Onset temperature, Tp=Peak temperature, Te=End temperature, ΔH=The enthalpy of the transition.

Samples	To (°C)	Tp (°C)	ΔH (J/g)
Yoghurt cakes			
Control (1)	72.64	74.44	0.887
S1	69.65	69.81	0.880
Sponge cakes			
Control (2)	70.88	72.66	0.877
S2	68.87	70.97	0.907

3.5 Water activity

Water activity is an important consideration for food product design and food safety, where food designers use water activity to formulate shelf-stable food. Mean water activity of control cake samples (1 & 2) and their low calorie cakes products (S1 & S2) were determined and presented in Table 6. Water activity of yoghurt cake (control 1) and sponge cake (control 2) were 0.92 and 0.89, respectively. While low calorie products of yoghurt cake (1) and sponge cake (S2) decreased to 0.72 and 0.81, respectively. This result indicates that the use of sucralose instead of sucrose in yoghurt or sponge cakes able to reduce water activity of cake product. Consequently, using sucralose instead of sucrose able was to improve the shelf-life of yoghurt or sponge cakes. This result agreed with Zoulias [27] who reported that acesulfame-K as a replacement for sucrose able to decrease the water activity of cookies. Also, Wetzel [28] stated that, TruviaTM as replacement for sucrose in doughnuts product reduced its water activity compared to those prepared with sucrose. Furthermore, Kim [29] reported that water activity was decreased as the concentration of polyols increased during replacing sucrose in formulation.

3.6 Sensory evaluation

The sensory properties (Cells, grain, Texture, Crumb color and flavor) of yoghurt cake (control 1) and sponge cake (control 2) was compared with their low calorie prepared cakes (S1 & S2). Table 7 indicated that, grain, Texture and flavor of cakes samples were not affected significantly in case of using sucralose in yoghurt cake (S1) or sponge cake (S2) compared to control cakes (1 & 2). But a significant difference in Cells and crumb color was observed in S1 and S2 cake samples if compared with control cake samples.

Table 6. Water activity and water content of yoghurt cake (control 1), sponge cake (control 2) and their low-calories cakes products, Mean values ± Standard Deviation, S1=Yoghurt cakes of replaced sucrose by sucralose, S2=Sponge cakes of replaced sucrose by sucralose.

Samples	**Water activity (aw)**	**Water content (%)**
Yoghurt cakes		
Control (1)	0.92 ± 0.03	25.7 ± 0.15
S1	0.72 ± 0.05	21.0 ± 0.19
Sponge cakes		
Control (2)	0.89 ± 0.04	28.0 ± 0.14
S2	0.81 ± 0.0.2	22.0 ± 0.12

Table 7. Sensory properties of yoghurt cakes (control 1), sponge cakes (control 2) and their low-calories cakes products, Mean values ± Standard

Deviation, S1=Yoghurt cakes of replaced sucrose by sucralose, S2=Sponge cakes of replaced sucrose by sucralose, LSD=Lest Significance Difference, NS=Non Significant difference.

Samples	Cells (30)	Grain (16)	Texture (34)	Crumb color (10)	Flavor (10)
Yoghurt cakes					
Control (1)	24.29b ± 2.82	14.13 ± 1.01	30.17 ± 2.17	8.71b ± 0.68	9.15 ± 0.6
S1	21.81c ± 3.77	13.11 ± 1.06	27.35 ± 2.38	7.82c ± 1.20	8.33 ± 1.09
Sponge cakes					
Control (2)	26.11a ± 2.20	13.25 ± 0.33	28.14 ± 2.26	9.65a ± 1.03	8.61 ± 1.19
S2	22.88c ± 1.13	13.02 ± 0.19	25.24 ± 3.15	7.35c ± 0.96	7.65 ± 0.95
LSD at 0.05	2.55	NS	NS	1.04	NS

3.7 Baking quality and freshness

The effect of replacing sucralose instead of sucrose on baking quality and freshness of yoghurt and sponge cakes were studied. Table 8 indicated that, weight and cake volume of low calorie cakes of yoghurt cake (S1) or sponge cake (S2) was decreased significantly compared to control cakes (1 & 2), while, specific volume was increased. The specific volume of baked cake indicates the amount of air that can remain in the final product. The higher gas retention and higher expansion of the product leads to a higher specific volume [30] After baking at 180°C for 35 min, the specific volumes of yoghurt cake (S1) and sponge cake (S2) cake were significantly higher than cakes of control (1 & 2). This result indicated that using sucralose instead of sucrose to prepare yoghurt or sponge cakes was able to remain a higher amount of air in cake samples [31,32].

The effect of storage period for 7 days at room temperature on freshness of low calorie yoghurt and sponge cakes products was evaluated. Table 8 showed that, yoghurt cake control of sucrose had highest alkaline water retention capacity, where it was declined during 0,3 and 7 days of storage to 400, 360 and 320%, respectively. However, low calorie yoghurt or sponge cakes of sucralose caused a noticeable decrease in alkaline water retention capacity values at the same storage period. Such effect might be related to the replacement sucrose with sucralose.

4. CONCLUSION

From the obtained results it could be concluded that, sucralose is a suitable sweetener for low-calorie sponge or yoghurt cakes (sugar-free bakery products). Incorporating this formula should yield a high quality cake free from sucrose and shortening. Due to its low fat and low calorie content, consumption sponge or yoghurt cake of sucralose could prevent from risk of health problems such as coronary diseases, high blood pressure and cholesterol, diabetes and obesity.

Table 8. Freshness properties and baking quality of yoghurt cakes, sponge cakes and their low-calories cakes. Mean values ± Standard Deviation, S1=Yoghurt cakes of replaced sucrose by sucralose, S2=Sponge cakes of replaced sucrose by sucralose.

Samples	Baking quality			Water retention capacity (Freshness)		
	Weight (g)	Volume (cc)	Specific volume	Zero time	3 days	7 days
Yoghurt cakes						
Control	130 ± 1.10	242 ± 1.68	1.82 ± 0.11	400 ± 2.02	360 ± 1.65	320 ± 1.16
S1	120 ± 1.15	235 ± 1.72	2.4 ± 0.09	360 ± 2.09	310 ± 1.44	270 ± 1.22
Sponge cakes						
Control	135 ± 1.20	284 ± 1.79	2.18 ± 0.15	370 ± 2.16	320 ± 1.10	260 ± 1.30
S2	125 ± 1.30	250 ± 1.82	2.64 ± 0.08	320	270 ± 1.19	220 ± 1.40

REFERENCES

1. Knecht RL (1990) Properties of Sugar. In: Pennington NL, Baker CW (Ed)Sugar. Van Nostrand Reinhold, New York, pp. 46-65.

2. Newsome R (1993) Sugar Substitutes. In: Altschull AM (Ed) Low-calorie Food Handbook. Marcel Dekker, New York, pp. 139-170

3. Abdullah A, Cheng TC (2001) Optimization of reduced calorie tropical mixed fruits Jam. Food Qual Prefer 12: 63-68.

4. Hess A, Setser CS (1983) Alternative systems for sweetening layer cake using aspartame with and without fructose. Cereal Chem60: 337-341.

5. Attia E, Shehata HA, Askar A (1993) An alternative formula for the sweetening of reduced-calorie cakes. Food Chem 48: 69-72.

6. Gautam A, Jha A, Singh R (2012) Sensory and textural properties of chhana kheer made with three artificial sweeteners. Int J Dairy Technol 66:109-118.

7. Lin SD, Lee CC (2005) Qualities of chiffon cake prepared with indigestible dextrin and sucralose as replacement for sucrose. Cereal Chem 82:405-413.

8. Kroger M, Meister K, Kava R (2006) Low-calorie sweeteners and other sugar substitutes: A review of the safety issues. Comprehensive reviews in food science and food safety 5:35-47.

9. Martinez-Cervera S, Sanz T, Salvador A, Fiszman SM (2012) Rheological, textural and sensorial properties of low-sucrose muffins reformulated with sucralose/polydextrose. LWT - Food Science and Technology. 45:213-220.

10. Chapello W (1998) The use of sucralose in baked goods and mixes. Cereal Foods World 43:716-717.

11. Giese JH (1993) Alternative sweeteners and bulking agents. Food Technol. 47:113-115,118,120-122,124-126.

12. Wallis KJ (1993) Sucralose: Features and benefits. Food Aus 45: 578-580.

13. Barndt RL, Jackson G (1990) Stability of sucrose in baked goods. Food Technology 44: 62-66.

14. Akesowan A (2009) Quality of reduced-fat chiffon cakes prepared with erythritol-sucralose as replacement for sugar. Pak J Nutr 8:1383-1386.

15. Maria KK Mohammad H, Hooman M (2015) Investigating the Effect of Different Sweeteners on the Process of Staling and Hardness of Packaged Sponge Cakes during Preservation Period. IJACS.

16. Bennion EB, Bamford GST (1997) The technology of cake making. Published by blacking academic and professional, Chapman & hall, London(6thedn) pp. 112-120,277,285-288.

17. AACC (2000) Approved Method of the AACC. 10th ed., American Association of Cereal Chemists, INC. st., Paul, Minnesota, USA.

18. Yamazaki WT(1953) An alkaline water retention capacity test for the evaluation of cookie baking potentialities of soft winter wheat flour. Cereal Chemistry 30: 121-246.

19. Kitterman JS, Rubenthaler GL(1971) Assessing the quality of early generation of wheat selections with the micro AWRC test. Cereal Sci. Today16: 313-316,328.

20. McClave JT, Benson PG (1991) Statistics for business and economics. Maxwell Macmillan International Editions. Dellen Publishing Co.USA 272-295.

21. Kim YS, Ha TY, Lee SH, Lee HY (1997) Effect of rice bran dietary fiber on flour rheology and quality of wet noodles. Korean. JFST 20: 90-95.

22. Kordonowy RK, Young VL (1985) Utilization of durum bran and its effect of spaghetti. Cereal Chemistry 62: 301-306.

23. Ramy A, Salama MF, Shouk AA (2002) Pollards a potential source of dietary fiber for pasta manufacture. EJFS 30: 313-330.

24. Bourne MC (2002) Food Texture and Viscosity. Concept and Measureme nt, (2ndedn) Academic Press. London p. 427.

25. Karaoglu MM, Kotancilar HG (2009) Quality and textural behaviour of par-baked and rebaked cake during prolonged storage.

26. Lim H, Setser CS, Paukstelis JV, Sobczynska D (1992) 17O nuclear magnetic resonance studies on wheat starch-sucrose-water interactions with increasing temperature. Cereal Chemistry 69:382-386.

27. Zoulias E, Piknis S, Oreopoulou V (2000) Effect of sugar replacement by polyols and acesulfame-K on properties of low-fat cookies. JSFA80:2049-2056.

28. Wetzel E (2010) The effect of various alternative sweeteners has on texture, color and water activity on fried doughnuts. Purdue University.

29. Kim Y, Lee J, Kim H, Choi SJ, Shin WS, et al. (2003) Sensory and physicochemical properties of selected sweetener blends containing polyols. Food Science and Biotechnology 12:151-156.

30. Gomez M, Oliete B, Rosell CM, Pando V, Fernandez A (2008) Studies on cake quality made of wheat-chickpea flour blends. LWT Food Sci. Technol 41: 1701-1709.

31. AOAC (2000) Official Methods of Analysis, (17thedn) Association of Official Analytical Chemists, Gaithersburg. MD.

32. Fung TT, Hu FB, Pereira MA, Liu S, Stampfer MJ, et al. (2002) Whole-grain intake and the risk of type 2 diabetes: a prospective study in men. Am J Clin Nutr 76: 535-540.

33. Hussein AMS, Abd El-Azeem AS, Hegazy AM, Afifi AA, Ragab GH (2011) Physiochemical, sensory and nutritional properties of corn-fenugreek flour composite biscuits. AJBAS5: 84-95.

10

THE MEDITERRANEAN DIET BETWEEN TRADITIONAL FOODS AND HUMAN HEALTH: THE CULINARY EXAMPLE OF PUGLIA (SOUTHERN ITALY)

Massimiliano Renna[a,], Vito Antonio Rinaldi[b], Maria Gonnella[a]

[a] Institute of Sciences of Food Production, CNR—National Research Council of Italy, Via G. Amendola, 122/O, 70126 Bari, Italy

[b] Vocational School for Hotel Hospitality and Gastronomy "A. Consoli", Contrada Rosatella, 70013 Castella Grotte (BA), Italy

ABSTRACT

The Mediterranean Diet, through a healthy profile of fat intake, carbohydrate at low glycaemic index, high content of dietary fibre, antioxidants and anti-inflammatory compounds, reduces the risk of certain pathologies such as cardiovascular disease. However, it represents anything much more than a dietary regimen and it is also characterised by its links to the various food cultures of the different countries of the Mediterranean area. On the other hand, the traditional cuisine of the Puglia region (Southern Italy) is based on a nutritional model mainly vegetarian because only a small share of calories is of animal origin; cereals are the basic ingredient, pulses and olive oil the main protein and fat source, respectively. In this paper we reported the culture, history, identity and heritage of this culinary model of the Southern Italy tradition to understand possible linkages with the

Mediterranean Diet. Moreover, some traditional recipes of the Puglia cuisine are examined as regards gastronomic and nutritional aspects, also for better explaining its relationship with the food style best famous in the world.

KEYWORDS

Lifestyle; History; Nutritional characteristics; Traditional cuisine; Puglia's dishes

1. MUCH MORE THAN A DIETARY REGIMEN

As explained by most dictionaries and shown by the media, the use of the term "diet" is frequently associated to specific food regimes, normally of a restrictive nature and with therapeutic or aesthetic purposes. Nevertheless, etymologically the word *dieta* comes from the Latin *diaeta* which in turn comes from the Greek *díaita*, "way of living", "life regime" (Corominas, 2000). Anatole Bally (a renowed Hellenist) in his dictionary (Bally, 2000) translates the term *díaita* as type of life and specifies as a first meaning *...in general, a whole set of habits of the body and the spirit, tastes, customs, etc...* Classical authors already used this term with the same sense as that used by Bally: Hippocrates in his works; Plato in *Republic* and *Laws*; Herodotus in *The Histories*; Pindar in *Pythian*; etc. This "way of living" is what we also know today as "lifestyle". Style or way of life that characterise individuals, groups, communities or villages and which make them similar or different from each other. Therefore, the Mediterranean Diet is this evidently dynamic and ever changing lifestyle with all its diversity, which it is possible to consider in its holistic sense made from tangible and intangible aspects and values. Other plural designations such as Mediterranean Diets and Mediterranean cuisines are the result of an argument used by those who believe that we cannot talk about "one" single diet, but that there are "many" diets in the Mediterranean just as there are many cuisines. It is a persistent argument which is also applied to the Mediterranean by those who think it is not relevant to consider it as a whole.

The Mediterranean Diet as an unequalled legacy of landscapes, places, knowledge, know-how, technologies, products, myths and beliefs, accents, creativity and hospitalities. It is in short, a (or perhaps "the") common language of the Mediterranean people. Furthermore, another important milestone in the dissemination of its significance was been the inscription of the Mediterranean Diet in the List of Intangible Cultural Heritage of Humanity by UNESCO, in November 2010. In no case this recognition represents an ultimate milestone, but rather an undertaking to work to safeguard this legacy.

2. A COMBINATION OF HISTORY, CULTURES AND ENVIRONMENT

The traditional Mediterranean Diet is the heritage of millennia of exchanges amongst the peoples and cultures of the Mediterranean Basin. It formed the basis of eating habits throughout the region until the mid twentieth century, but it is now gradually being lost due to the spread of the western-type economy and urban and technological society as well as the globalisation of production and consumption.

The Mediterranean culture integrates the past and the present, because many of its modern aspects can be traced to the ancient past. Effectively, the Mediterranean basin has been, for millennia, a crossroads of civilisations as a point of convergence for people, merchandise and religions. In Tunisia, for example, you can find Portuguese and Spanish survivals in the consumption of "*bacalao*" or salt cod. It was probably the Portuguese who introduced the cod into Tunisia, as shown by the adoption of the Arab name "*baqalaw*". It is now mainly consumed in a porridge, chiefly in Sfax, on the feast of*Aïd el Kebir*, to accompany a sweet and sour dish, the ritual "*charmoula*". Inhabitants of the inland areas tended rather to eat a mollusc dried in the sun and salted, called "*ouzef*".

Even the scenery represents, in some ways, a composite picture: some plants such as olive trees, wheat, and vineyards exist in this area from time immemorial. Oranges and lemons, however, originate from the Far East and were brought to the Mediterranean by Arabs. Thorny plants such as prickly pear were brought from America. Also tomatoes, eggplants, corn and potato have been imported, but these and several other plants have gradually become integral parts of the Mediterranean diet. Indeed, the benevolent climate in the region and the human activities have moulded disparate components into a balanced ecosystem (Braudel et al., 1985).

As regards the religion, during polytheistic and monotheistic times the influence of the sacred and the liturgical on diets, production and habits has been substantial. Amongst the celebrations and deity festivities, the *yumuaa*, the *shabbat*, the *dominica*, saints' feasts and other holy days of obligation, the Mediterranean has spent half its life offering up or praying... or disputing the monopoly of the divine truth. Foods here have been raised to the category of the sacred. Bread, wine and oil (the Mediterranean triad) occupy a prominent place: oils anoint the newborn, the king and the moribund; Athena's olive tree founds the state-city and represents the most useful offering for humankind; the dove with the olive branch announces the end of the Flood; the olive of the Koran is the "the blessed tree which is neither from the East nor the West and whose oil seems to take light without coming into contact with fire"; the wine and bread at the Last Supper, are the body and blood of Jesus Christ. The list is endless, in all peoples and times of the Mediterranean. Today we have almost forgotten that when we refer to a companion we are referring to a *cum panis*, someone with whom we are sharing bread.

Conceived as a constantly evolving lifestyle, the Mediterranean Diet is a complex system of shared knowledge relating to health, food, cultures and people; it is the product of a particular environment, a geographical region of multiple facets and rich history, which conserves traditional knowledge and a diversity of foods and diets.

The shortage of resources has prompted communication amongst Mediterraneans and the construction of a food style that has adapted and transformed plants coming from remote places. Their food habits are excellent examples of the fertile combination of production and commercial activities that has shaped the history of mankind. Thus, few Mediterranean dishes can be imagined without such exchange. If we think about some of the most famous examples, it is evident that they have only existed thanks to the Mediterraneans' eagerness to broaden the horizons of their diet. This is the case of the Greek *moussaka* (aubergines and meat), the Maghrebi *harira* (tomato and vegetable soup), the *hummus,* a dish shared by Arabs and Jews (chickpea and sesame spread), the Egyptian *falafel* (bean fritters), the Turkish *dolmathakia* (rice-stuffed grape leaves), the Italian *pesto* (basil sauce), the French *bouillabaisse* (fish and tomato stew) or the Spanish *pisto* (vegetable *ratatouille*) and rice dishes. Their names recall the taste of Mediterranean cuisines; but this taste is the result of combining ingredients and spices that did not exist until they reached, at a given moment in history, the big larder of the Mediterranean and became part of its long history of adaptation (Abdelkader, 1998 and Chomski, 2010).

In conclusion, the Mediterranean has been a permanent laboratory for mankind to show the ability to adapt and disseminate foods. Anyway, the inhabitants of the Mediterranean have built their food style through interaction or by defining their own habits as opposed to those coming from the other side of the basin. Thus, interaction and opposition are the two complementary sides of Mediterranean cultures and also of their diets.

3. NUTRITIONAL AND HEALTHY CHARACTERISTICS OF THE MEDITERRANEAN LIFESTYLE

In 1975 the American physiologist Ancel Keys published, together with his wife Margaret, the book "How to Eat well and Stay well". The Mediterranean Way, which talks about the diet of common people in Greece, Italy and along the Mediterranean coasts of France and Spain (Keys and Keys, 1975). Known as the Seven Countries Study, it was the result of more than 25 years of cross-national studies on the subject of diet–cholesterol–coronary diseases and became a bestseller. Ancel Keys can be considered the discoverer from the scientific evidence, of the Mediterranean Diet and its benefits. Through his works and publications, but also thanks to his sensitivity, his observation skills and his curiosity, the researcher Ancel Keys not only showed a transcendental relationship between diet and cardiovascular diseases but also discovered and disseminated with passion a lifestyle that fascinated him and which he understood as indivisible and substantial to everything he was studying in the Mediterranean. His interdisciplinary approach, which has been adopted lately through sociology, anthropology, economics, agronomy and biology, among others, has helped to extend its perception considerably. On the other hand, as previously reported, the Mediterranean is characterised by various food models influenced by different cultures, histories and religions. Nevertheless, although different regions in the Mediterranean basin have their food model, it is appropriate to consider these as variants of a single entity, the Mediterranean Diet. Indeed, the dietary patterns that prevail in the Mediterranean have many common characteristics, most of which stem from the fact that olive oil occupies a central position in all of them. Olive oil is important not only for its own health benefits, but it is also associated with the consumption of large quantities of vegetables in the form of salads and equally large quantities as cooked foods. Other essential components of the Mediterranean diet are cereals (especially the wheat) and their derivative products, the fish and shellfish, legumes and nuts. Total fat may be high, around or in excess of 40% of total energy intake (as in Greece), or moderate, around 30% of total energy intake (as in Italy). In all instances, however, the ratio of monounsaturated/saturated fats is much higher than in other regions of the world, including northern Europe and North America. Thus, it might be convenient to define the Mediterranean Diet as a plant-based pattern, where vegetables, fruits, cereals (preferably as whole grain), legumes and nuts should be consumed in high amount and frequency. This pattern also includes moderate consumption of fish (especially the blue-fish), white meat, eggs, milk and dairy products, principally yogurt and cheese (Willet et al., 1995). On the contrary, consumption of red meat, processed meats, and foods rich in sugars and saturated fats should be small in both quantity and frequency. An adequate daily intake of water should be guaranteed, as well as moderate consumption of wine (preferably the red type) is recommended. In this context, it is important to specify as the consumption of wine, especially during family meals, is due to social and religious aspects (Willet et al., 1995). On the other hand, the high consumption of vegetables, fresh fruits and whole cereals as well as the habitual use of extra virgin olive oil guarantee a high intake of monounsaturated fatty acids, carotenoids, ascorbic acid, tocopherols, various vitamins, mineral and several healthy substances, such as polyphenols and anthocyanines. The vegetable products are very important also for their intake of dietary fibre, which improves intestinal peristalsis and reduces the glycaemic index of the meal. Instead, the low consumption of red meat and its derivates, together with the moderate consumption of milk and dairy products allows to reduce the intake of the exogenous cholesterol and saturated fatty acids. On the other hand, the combined use of pulses and cereals allows to ingest vegetable proteins with high nutritional value, while the blue-fish guarantees an adequate intake of polyunsaturated fat acids. High consumption of dietary fibre, low glycaemic index and glycaemic load, anti-inflammatory effects, and antioxidant compounds may act together to produce favourable effects on health status (Visioli and Galli, 2001, Pitsavos et al., 2005, Estruch et al., 2009 and Estruch et al., 2006). Moreover, the Mediterranean Diet is associated with a lower incidence of mortality and is also related to lower incidence of cardiovascular diseases, type 2 diabetes, certain types of cancer and neurodegenerative diseases (Estruch et al., 2013, Mitrou et al., 2007, Salas-Salvadó et al., 2013, Sofi et al., 2010 and Sofi et al., 2013). However, people from Mediterranean countries are changing the traditional Mediterranean Diet and include low nutrient dense foods (such as sugared soft drinks, sweets, bakery products, salted snacks) or vary their food processing methods (such as refinement of flour) towards a less healthy diet. These changes may have contributed to an increased risk of deficient intakes for some vitamins, especially folates, vitamins A and D, as well as inadequate intakes for the rest of the vitamins, in particular among certain population groups or collectives (Serra-Majem et al., 2002).

Finally, seasonality, biodiversity, the use of traditional and local food products are the key elements of the Mediterranean Diet for better understanding how all these differences may be translated in several "Mediterranean Diets". These are also characterised by qualitative, cultural and lifestyle elements, such as conviviality, culinary tradition and physical activity. For these reasons, in agreement with some Authors (Bally, 2000 and Corominas, 2000), could seem reductive to consider this pattern only as dietary models, while could be more appropriate to speak about the Mediterranean lifestyles.

4. BACKGROUND OF THE TRADITIONAL CUISINE IN PUGLIA

Puglia region has been, for centuries, a territory dominated by several dynasties that has given an inheritance of different customs and knowledge included culinary ones. The union of its historical experiences and variegated environmental characteristics of the territory has originated the traditional cuisine of Puglia, which is clearly linked to healthy principles of the ancient Mediterranean Diet. In fact, in this cuisine great space is reserved to extra virgin olive oil, vegetables, sea products and cereals. The use of these products as ingredients to prepare traditional dishes has been due to their availability in a variegated territory. Therefore, it is important to underline that the Puglia region is relatively long (350 km) and narrow (60 km) with an area of about 19,358 km^2. This area, open to Adriatic and Ionian Seas with a coastal zone of nearly 800 km, is prevalently level to slightly sloping, with more than 60% of territory below 200 m above sea level (Elia and Santamaria, 2013). Puglia is located in the southern part of Italy and it is possible to place this region in the centre of the Mediterranean basin. Moreover, the geographical length and the variety of orographic and pedoclimatic conditions give rise to the numerical and quantitative richness of vegetables grown in the region (Elia and Santamaria, 2013).

Puglia, which has a long tradition in vegetable growing, is particularly rich in local vegetable varieties, obtained by farmers themselves after repeated simple selection procedures generation after generation (Elia and Santamaria, 2013). Only in this Italian region a relevant number of traditional dishes can be found based on pasta and vegetables as main ingredients, which cannot be found in such number in other Italian regions. Therefore, several local varieties of vegetables for which there is a strong link with the regional traditions are used as ingredients to prepare traditional dishes of the Puglia's cuisine. On the other hand, these local vegetables are very appreciated both as refined food and for the intake of several healthy nutrients. For example, the fruits of*carosello* and *barattiere* (herbaceous plants belonging to *Cucumis melo* L. species) (Fig. 1) are consumed at the immature stage, fresh and raw, instead of cucumbers, due to their better quality profile. They are characterised by being refreshing and digestible as well as having high potassium and low reducing sugar and sodium contents (Serio et al., 2005). The *carota di Polignano a Mare* (a multicoloured landrace of *Daucus carota* L.) (Fig. 2) is greatly appreciated by local people for its special taste, tenderness, crispness, flavour, fragrance and great variety of colours, that range from yellow to dark purple in the outer core and from pale yellow to light green in the inner core (Cefola et al., 2012). It is used as a raw vegetable or as an ingredient in a wide range of dishes; from appetisers to desserts, the *carota di Polignano a Mare* shows great potential for culinary applications. Moreover, for its high content in antioxidant activity, total phenols, carotenoids and β-carotene it can be regarded also as a functional food. The interesting traits of other local vegetables used in some traditional recipes are better reported in the following paragraphs, however it is important to underline that a particular segment of typical Puglia's vegetables is represented by wild edible plants (WEP), which includes some progenitors of cultivated vegetables. The food use of WEP is deeply rooted in the tradition of the region and they are part of many regional recipes. For example in the Foggia cuisine, WEP alone or in combination with pasta, bread (*pancotto*) or pulses, are always present on local tables. The historical figure of *terrazzani*, who are collectors and sellers of wild herbs, mushrooms, frogs, snails and *lampascioni*, is typical in Capitanata (Foggia province). Still today examples of this singular figure of *ante litteram* ecologist can be found at the local markets in Foggia and its province with their rich load of spontaneous biodiversity collected in the woods and the neighbouring rural areas. The following are significant examples for Puglia: wild cardoon, wild asparagus, rocket, wild chicory, purslane, some leaf beets and many other species (Bianco et al., 2009,Gonnella et al., 2010 and Elia and Santamaria, 2013). Bianco et al. (2009) reports a list of 532 spontaneous species present in the

Puglia region which have been used in the past and until recently as food, indicating the botanical family and name, common name, frequency in the territory, the edible part and how it is prepared. Some of these species have been subjected to more or less successful attempts of cultivation. The cases of*lampascione*, wild asparagus, wild cardoon, borage, wild fennel, sage and rosemary are indicative in this respect. Among these examples of domestication, one of the most successful is that of wild rocket (*Diplotaxis tenuifolia* (L.) DC.) (Fig. 3), which, since the first attempts in Italy in early 1990s, is currently grown over a 1000 ha area, mainly in greenhouses. Since 1990 there has been an increase in its consumption also due to the initiatives of some great chefs who proposed rocket in various recipes. Thus, for example, rocket is used as an ingredient in *risotto*, *gnocchi*, *tortellini*, different types of pasta, pizzas, *piadine*, with canapés and *bruschetta* and to prepare *carpaccio*. In addition, it is used by small scale artisan processors to prepare creams, sauces and pesto.

Figure. 1. Different Puglia's local varieties of *carosello* and *barattiere* (*Cucumis melo* L.).

Figure. 2. Assortment of Polignano carrots: it is interesting to note the great variety of the root colours which is highly appreciated by consumers.

Figure. 3. Rocket on pizza after cooking.

Another important trait of the traditional cuisine of Puglia is the predominant role that have cereal-based products, especially wheat. This custom is undoubtedly linked to the local habit which involves the constant presence of the bread on the table in all its many regional varieties, or the pasta of durum wheat as well as the soups that include several type of whole cereals. *Cavatelli, orecchiette, lagane, maccaruni* are the name of some type of traditional pasta in Puglia. Interesting is the case of the "*pasta di grano arso*" which is made from wheat that is toasted and then crushed. Pasta made from this type of flour tends to be purple brown. Originally, this type of wheat was obtained from gleaning the wheat fields after the stubble was burnt, when the left-over spikes from the harvest, burned in the fire, were collected. Many types of traditional pasta, bread and bakery products represent the principal ingredients of several Puglia's recipes and are produced by the use of local varieties of wheat and the application of traditional processing. Some of these product have obtained also the Denomination of Protected Origin (DPO). It is the quality mark that is assigned by the European Union to those food products that have unique and unmistakable features that depend exclusively, or mainly, on the area where they are produced. The geographic location includes both the natural (climate, environment) and human (traditional production techniques, craftsmanship) factors which are combined to produce something that cannot be imitated outside the specific production zone. For a product with the DPO, the production, transformation and processing phases must all take place within a specific geographic area. The conditions to achieve EU recognition must be listed in a specific protocol, with the production method, the product name and description, the geographic confines and all the limitations and prescriptions relative to the product production and labelling method. For example, "*pane di Altamura* DPO" (bread of Altamura city) is a bakery product made from durum wheat flour, according to the ancient method of production which requires the use of starter dough, sea salt and water. The starter dough has always been a very important aspect for making traditional bakery products in Puglia as well as in other Mediterranean countries. In fact, in the past it was the only type of yeast, while nowadays it is used to obtain products with best sensorial traits. The processed flour used to prepare "*pane di Altamura* DPO" is obtained from the grinding of various types of durum wheat: "appulo", "arcangelo", "duilio" and "simeto" which are produced in the area where the bread is to be made. Bread is always been a staple diet of the population of Puglia. Large loaves used to be prepared at home and taken to public wood-burning furnaces for baking. Another type of traditional bread is *pane di Laterza* also prepared by the use of durum wheat flour made from local varieties of wheat. The first reports relating to the production of bread in this district of Puglia dates back to the 5th century B.C. The bread, at the time known as*focaccia*, was made from spelt and barley, with or without yeast. It was cooked under ashes or on embers. It is important to emphasise that in the past the flour was produced with millstones (Fig. 4), allowing to obtain a whole wheat flour rich in fibre with a lower glycaemic index than present refined flours.

Figure. 4. Old millstone used in the past to produce flour.

Finally, the virgin olive oil represents a very important ingredient of Puglia's cuisine not only for its health properties. In fact, it contains a great number of volatile and non-volatile compounds that enrich the taste of many dishes by its fragrant and peculiar flavours. In this context the complex flavour of virgin olive oil by fruits of cultivar from Puglia is appreciated internationally by gourmets and cherished by native consumers. There are different virgin olive oils with DPO in Puglia, and each has different characteristics and traditions. For example "*Terra di Bari* DPO" oil is yellow to green in colour, and has a fruity flavour with a slight piquant and bitter taste. The smell is redolent of the fragrance of grass and fresh almonds. The olive growing tradition is very strong in the province of Bari. It was already a thriving trade in Roman times. It subsequently gained momentum in medieval times and spread as far Venice from where it reached other countries of Continental Europe. On the other hand, "*Terra d'Otranto* DPO" oil gets its name from the word formerly used by Basilian monks to indicate the area within the arc of Murge as well as within the Ionian and Adriatic coasts. In this case the cultivation of the fruit was introduced by the Phoenicians and Greeks, but it was the Basilian monks who made it a thriving activity. This type of virgin olive oil presents a yellow or green colour with greenish tint. Fruity taste that is slightly piquant and bitter and fruity fragrance with hints of leaves are also typical in "*Terra d'Otranto* DPO" virgin olive oil. Other typical olive oil in Puglia are "*Dauno* DPO", "*Collina di Brindisi* DPO" and "*Terre Tarentine* DPO" to be tasted with vegetables and pulses as well as with *cruditè* and fish. Puglia region with about 40 millions olive trees must only have so many wonderful virgin olive oils (Puglia Region, 2010).

Finally, it is important to underline as the food products described in this paper represent only a piece of all the variety and richness of the typical food production made in Puglia. Other several examples could be reported, nevertheless making a complete list of food products from all Puglia areas was not the aim of this work.

5. PUGLIA'S DISHES: BEYOND A SIMPLE EXPERIENCE OF TASTING

Puglia is a land with a traditional cuisine in which sea products, vegetables and animal foods are well mixed in several recipes.

Therefore, the taste and flavours of sea and countryside can combine itself in a dish to create complex sensorial perceptions but always intriguing. On the other hand, the sensorial experience of these dishes is only one of the positive reasons to consume them. In fact, many of ingredients used to prepare Puglia's dishes have nutritional and healthy characteristics. In this context it is important to underline that some essential ingredients of the Mediterranean diet (such as cereals and their derivates, fish, legumes and olive oil) have been the criteria used for selecting the recipes. Thus, some of the recipes that characterise the traditional cuisine of Puglia as well as healthy benefits of its ingredients are showed in the following dishes. Anyway, two important aspects could be considered: (1) not always a recipe can be considered diffused in all areas of the region; (2) Puglia has a very articulated variation of typical dishes throughout all the territory.

5.1 Fave bianche e cicorie

In this dish a purée of husked broad beans (*Vicia faba* L.) cooked in earthenware is presented with boiled vegetables and raw *Cipolla di Acquaviva delle Fonti* (a local variety of *Allium cepa* L.) (Fig. 5). The extra virgin olive oil, poured at the end of preparation, represents not only the seasoning but also a fat source which is rich in monounsaturated fatty acids. The origin of this dish is very old, and probably it is the result of the influence of different cultures, since a similar recipe (i.e. Greek Fava) represents a traditional dish of the Greek cuisine. Anyway, *fave bianche e cicorie* is a very nutritious dish since broad beans are high in protein and carbohydrates with low glycaemic index. Moreover, the local habit to consume also bread together this dish improves the nutritional quality of the proteins as regards the essential amino acids content. In the traditional recipe, dried broad beans without the husk (the leathery outer cuticle) are soaked for about 10 h and then, cooked with water and some spice until a purée is obtained. The term *cicorie* is traditionally used to indicate WEP (i.e. *Chicorium intybus* var. *silvestre* Bischoff.,*Sonchus oleraceus* L., *Helminthoteca echioides* L., etc.), however it is possible to use different types of cultivated chicory (*C. intybus* var. *foliosum*) as single vegetable or mixed with WEP. Finally, the use of *Cipolla di Acquaviva delle Fonti* allows to improve the chromatic aspect of this dish as well as its taste. In fact, this local variety of onion is renowned for its sweetness and is recognised by its colour between carmine red and purple, that becomes lighter toward the inside, until it becomes completely white.

Figure. 5. Fave bianche e cicorie*: purée of husked broad beans presented with boiled vegetables and raw* Cipolla di Acquaviva delle Fonti.

5.2 Orecchiette e cime di rapa

For this traditional recipe (Fig. 6) the most famous fresh pasta of Puglia (*orecchiette*) is mixed with broccoli raab (so-called *cima di rapa*). *Orecchiette* are produced by mixing durum wheat flour and water and are made in the typical round shape. According to Sada (1991)*orecchiette* were found in Puglia in the second half of the XVI century. It would seem that their origin is in Sannicandro di Bari (a little town of central Puglia), while the name is derived from a Jewish sweet: "*orecchie di Aman*" (Aman's ears). *Cima di rapa*(*Brassica rapa* L. subsp. *rapa sylvestris* var. *esculenta*) is an ancient species of Mediterranean origin, linked to the culinary traditions of a large part of Central-Southern Italy. This vegetable is mainly cultivated in Puglia, where, since July 2006, it has been included in the list of regional traditional products. In the last decades, moreover, this vegetable has attracted the attention of an increasing number of consumers in Europe, United States, Canada, Argentina and Australia. In North America *cime di rapa* are also known by the names of *broccoli rabe, broccoli de rabe, raab, rapa, rappini, rapini, spring broccoli, Italian turnip and taitcat* (Di Gioia and Santamaria, 2009). The popularity of this typical Puglia's vegetable is probably due both to its aromatic taste and content of glucosinolates, well-known as important healthy compounds. After tasting of this dish is possible to perceive a little bitter sensation from *cime di rapa*. However, the overall taste is well balanced by organoleptic characteristics of both *orecchiette*, extra virgin olive oil and eventually hot chilly pepper and a few anchovy fillets.

Figure. 6. *Orecchiette e cime di rapa*: fresh Puglia's pasta with broccoli raab.

5.3 Soup of pulses and whole wheat

This is an important traditional dish (Fig. 7) which was the basic food of many families in the past. Effectively, legumes and grain self-produced made it possible to satisfy both the energy needs and protein intake, so the pulses were also considered as the "poor man's meat". To prepare this dish several traditional pulses of Puglia are used. For example*cicerchia* (*Lathyrus sativus* L.) which is an ancient pulse similar to the chickpea, with grains that are wedge-shaped, sharp, whitish, grayish-brown, or yellowish in colour. Bean (*Phaseulus vulgaris* L.) of "Monti Dauni Meridionali" has pleasant taste, with good digestibility and is easy to cook because of its thin skin. Also lentil (*Lens culinaris* L.) of "Altamura", pea (*Pisum sativum* L.) "nano di Zollino", fava bean (*Vicia faba* L.) of "Carpino" and black garbanzo bean are traditional pulses used together with whole wheat to prepare this main dish.

Figure. 7. *Zuppa di legumi*: several type of beans (fava bean, lentils, *cicerchia*, black garbanzo bean) and cereals.

5.4 Little savoury pie of anchovies

To prepare this dish (Fig. 8) the head of fresh anchovies is first removed, so they are opened in half, gutted and cleaned carefully. Then, the anchovies are placed in layers in a container for adding virgin olive oil, parsley, salt, pepper and bread crumbs. The filling used to stuff the pie is made up of eggs, cheese and fresh tomato. This is a typical example of a dish in which sea ingredients are combined with products from countryside to create a dish with very special flavours. The used blue fish also allows a good intake of essential polyunsaturated fatty acids. Also in this case it is possible to underline how the origin of the recipe has been influenced by other food cultures such as the Spanish cuisine.

Figure. 8. Little savoury pie of anchovies.

5.5 Salad of crenate broomrape

Crenate broomrape (*Orobanche crenata* Forsk.) also known as bean broomrape is a root parasite plant that produces devastating effects on many crop legumes and has become a limiting factor for faba bean production in the Mediterranean region. However, in Puglia it is considered a wild edible plant used to prepare several traditional dishes. On the other hand, crenate broomrape is interesting for its content of phenols that are both flavour components and healthy compounds as antioxidants. Traditionally, the stems are clean, washed and boiled in water and salt. After boiling, the stems are soaked in water to reduce the bitter taste and used as an ingredient in several recipes. In the present recipe crenate broomrape are used to prepare a salad with extra virgin olive oil, vinegar, mint and fresh garlic (Fig. 9). The sensory traits of this dish are complex but very intriguing. Indeed, sour and bitter taste are well combined with the pungency of the garlic and the freshness from the mint. Moreover, the lingering aftertaste and aromas make this a very tasty dish.

Figure. 9. Salad of crenate broomrape (*Orobanche crenata* Forsk.).

5.6 "Flowers" of *lampascioni*

Lampascioni is the common name used to indicate the bulbs of *Leopoldia comosa* L., that is a spontaneous plant growing in all

Mediterranean area. In Puglia the bulbs of this wild edible plant were used in traditional gastronomy for different preparations. Main tasks are: cleaning of bulbs by eliminating the roots and outer tunics, cross cutting the basal plant, washing, boiling and then soaking them in water to reduce their typical bitter taste. The product is typical of the tradition of the rural farms. Bruni (1857) indicated "*lampascioni sott'olio*" among the different ways to use the bulb. In this recipe*lampascioni* have been floured and fried in extra virgin olive oil for obtaining crispy snack that reminds a "flower" (Fig. 10). It is important to underline that the use of the extra virgin olive oil for frying means no harm for your health since it has a high smoke point. The taste of this dish is really intense and intriguing with a lingering aftertaste that results aromatic and slightly bitter. Therefore, the fried *lampascioni* are topped with *vincotto* that is a brown syrup obtained from grape must. The traditional method of producing *vincotto*requires 12–15 h of boiling freshly crushed grape must in a large copper or steel boiler; the grape are picked late to ensure a high sugar content. *Vincotto* is brown with more or less intense red reflection; it is very thick and sweet because of the sugar present in the must. In the past *vincotto* was so precious that it became the only condiment that the people used during the winter festivities and sometimes for therapeutic purpose.

Figure. 10. "Flower" of lampascioni (Leopoldia comosa L.).

6. CONCLUSIONS

In this paper is reported the culture, history, identity and heritage of the traditional cuisine of the Puglia region to understand possible linkages with the Mediterranean Diet. Effectively, the culinary model of this region of Southern Italy is mainly vegetarian because only a small share of calories is of animal origin; cereals are the basic ingredient, pulses and olive oil the main protein and fat source, respectively. However, in Puglia's cuisine sea products, vegetables and animal foods are well mixed in several recipes. Therefore, the taste and flavours of sea and countryside can combine itself in a dish to create complex sensorial perceptions but always intriguing. Nevertheless, the sensorial experience is only one of the positive reasons to consume these dishes, because many of the used ingredients have nutritional and healthy characteristics. Finally, knowledge of ancient recipes and culinary traditions may be an opportunity for innovation in gastronomy if new culinary techniques are applied in popular and typical dishes in order to enhance their sensory and nutritional properties.

ACKNOWLEDGEMENTS

The authors thank Dr. Pietro Santamaria and Dr. Angelo Signore for pictures number 1 and 2.

REFERENCES

1. Abdelkader, S.A., 1998. El Mediterráneo, de la integración a la fragmentación: los intercambios de la Antigüedad a nuestros días. Barcelona, Icaria.
2. Bally, A., 2000. Le Grand Bally Dictionnaire grec-français. Hachette Éduca- tion, Paris.

3. Braudel, F., Coarelli, F., Aymard, M., 1985. La Mediterranee: l'espace et l'histoire. Flammarion, Paris.

4. Bianco, V.V., Mariani, R., Santamaria, P., 2009. Piante spontanee nella cucina tradizionale molese. Levante Editori, Bari, Italy.

5. Bruni, A., 1857. Descrizione botanica delle campagne di Barletta. Stamperie e Cartiere di Fibreno, Napoli.

6. Cefola, M.,Pace, B.,Renna, M.,Santamaria, P.,Signore, A.,Serio, F.,2012. Composition alanalysis and antioxidant profile of yellow, orange and purple polignano carrots.Ital.J.FoodSci.24,284–291

7. Chomski, D., 2010. Cocinajudíaparacelebrarlavida. Comidasde fiesta, según fuentesclásicas, laCábalaylatradición. Somonte-Cenero,Gijón, Trea.

8. Corominas, J., 2000. Breve diccionario etimológico de la lengua española, 3rd ed. Madrid, Gredos

9. Di Gioia, F., Santamaria, P.,2009. Aimercati piacelacimadirapapugliese, ortaggio "antico". Ortofrutt. Ital.6,100–107.

10. Elia, A., Santamaria, P.,2013. Biodiversity in vegetablecrops, aheritageto save: the case of Pugliaregion.Ital.J.Agron.8,21–34.

11. Estruch, R.,Martínez-González,M.A.,Corella,D.,Basora-Gallisá,J.,Ruiz-Gutiérrez, V.,Covas,M.I.,Fiol,M.,Gómez-Gracia,E.,López-Sabater,M. C., Escoda,R.,etal.,2009.Effectsofdietary fibre intakeonriskfactors for cardiovasculardiseaseinsubjectsathighrisk.J.Epidemiol.Commu- nity Health63,582–588

12. Estruch, R., Martínez-González, M.A., Corella, D., Salas-Salvadó, J., Ruiz- Gutiérrez, V., Covas, M.I., Fiol, M., Gómez-Gracia, E., López-Sabater, M. C., Vinyoles, E., et al., 2006. Effects of a Mediterranean-style diet on cardiovascular risk factors: arandomizedtrial. Ann.Intern.Med.145, 1–11.

13. Estruch, R., Ros, E.,Salas-Salvadó, J.,Covas, M.I., Corella, D.,Arós, F., Gómez-Gracia, E.,Ruiz-Gutiérrez, V.,Fiol, M.,Lapetra, J., etal., 2013. Primary prevention of cardiovascular disease with a Mediterranean diet.N. Engl. J.Med.368,1279–1290.

14. Gonnella,M.,Charfeddine,M.,Conversa,G.,Santamaria,P.,2010.Purslan e: a reviewofitspotentialforhealthandagriculturalaspects.Eur.J.Plant Sci. Biotechnol.4,131–136.

15. Keys, A., Keys, M., 1975. How to Eat Well and Stay Well. The Mediterranean Way, Garden City (N.Y.), Doubleday.

16. Mitrou, P.N., Kipnis, V., Thiébaut, A.C.,Reedy, J.,Subar, A.F., Wirfält, E., Flood, A., Mouw, T.,Hollenbeck, A.R.,Leitzmann, M.F., etal., 2007. Mediterranean dietary pattern and prediction of all-cause mortality in a US population : results from the NIH-AARP Diet and Health Study. Arch. Intern. Med.167,2461–2468.

17. Pitsavos, C.,Panagiotakos,D.B.,Tzima,N.,Chrysohoou,C.,Economou,M., Zampelas,A.,Stefanadis,C.,2005.AdherencetotheMediterraneandietis associated withtotalantioxidantcapacityinhealthyadults:theATTICA study. Am.J.Clin.Nutr.82,694–699.

18. Puglia Region, 2010. Atlas of Typical Agri-Food Products of Puglia. Liantonio Editrice, Palo del Colle, Italy

19. Sada, L., 1991. La cucina della terra di Bari. F. Muzzo Editore, Padova.

20. Salas-Salvadó,J.,Bulló,M.,Estruch,R.,Ros,E.,Covas,M.I.,Ibarrola-Jurado, N., Corella, D.,Arós, F.,Gomez-Gracia, E.,Ruiz-Gutiérrez, V.,etal., 2013. Prevention of diabetes with Mediterranean diets as ubgroup analysis of arandomizedtrial.Ann.Intern.Med.160,1–10.

21. Serio, F.,Florio, G.,Parente, A.,Santamaria, P.,2005. Confronto trasistemidi coltivazione dicaroselloe barattiereinserra. II. Aspetti qualitativi.Colt. Protet. 34(Suppl.5),S28–S35.

22. Serra-Majem,L.,Ribas,L.,Pérez-Rodrigo,C.,García-Closas,R.,Peña-Quintana,L.,Aranceta,J.,2002.Determinantsofnutrientintakeamong

children andadolescents: results from the enKidStudy. Ann. Nutr.Metab. 46 (Suppl.1),31–38.

23. Sofi, F., Abbate, R., Gensini, G.F., Casini, A.,2010. Accruing evidence on benefits of adherence to the Mediterranean diet on health: an updated systematic review and meta-analysis. Am. J. Clin. Nutr. 92,1189–1196.

24. Sofi, F.,Macchi,C., Abbate, R.,Gensini, G.F.,Casini, A.,2013. Mediterranean diet and health. Biofactors 39,335–342.

25. Visioli, F.,Galli,C., 2001. The roleofantioxidants in the Mediterranean diet. Lipids 36,49–52.

26. Willet, W.C., Sacks, F.,Trichopoulou, A., Descher, G., Ferro-Luzzi, A., Helsing, E.,Trichopoulos, D.,1995. Mediterranean diet pyramid: acultural model forhealthyeating.Am.J.Clin.Nutr.61,1402S–1406S.

11

INTRODUCING DITTANY OF CRETE (ORIGANUM DICTAMNUS L.) TO GASTRONOMY: A NEW CULINARY CONCEPT FOR A TRADITIONALLY USED MEDICINAL PLANT

Nikos Krigas[a,,], Diamanto Lazari[b], Eleni Maloupa[c], Maria Stikoudi[c]

[a] Department of Botany, School of Biology, Aristotle University of Thessaloniki, GR-54124, Greece

[b] Department of Pharmacognosy, School of Pharmacy, Aristotle University of Thessaloniki, GR-54124, Greece

[c] Laboratory of Conservation and Evaluation of Native and Floricultural Species-Balkan Botanic Garden of Kroussia, Institute of Genetics, Breeding and Phytogenetic Resources, Hellenic Agricultural Organisation Demeter, GR-57001 Thermi, Thessaloniki, Greece

ABSTRACT

A new culinary concept has been developed to praise ancient and modern uses of exclusive Mediterranean ingredients, focusing the world's attention in a region: Crete, Greece. We reviewed the vernacular names, medicinal properties and traditional uses of the Dittany of Crete (local endemic of the island of Crete, Greece) and we explored the possibility of cooking different dishes. We developed a novel concept which resulted in the culinary use of the infusion, the leaves and/or inflorescences of this perennial herb in modern sweet and savoury dishes of Mediterranean cuisine (five case-studies are described and illustrated). Our study expands the use of a unique and beneficial herb (*Origanum dictamnus*) rendering it as a new spicy ingredient suitable for gastronomic experimentation. The promotion of new uses for this traditionally used medicinal plant (currently cultivated only at small scale on the island of Crete) (i) offers new ingredients to international gastronomy, (ii) may prove to be beneficial for local economies, and (iii) supports sustainable plant exploitation.

KEYWORDS

Infusion; Mediterranean dishes; Lamiaceae; Endemic; Threatened

1. INTRODUCTION

International gastronomy and food science are in search for appealing ingredients, new food products or new technology and methods for dish preparation. Different algae (blue, green, red algae) and super-foods (Morris, 2012) have entered the international cuisine due to their nutritional value and have found many applications which reflect an expanding market for related products (e.g. molecularg astronomy network.com, algaecompetition.com). Apart from delivering new original flavours and sensations, this new trend is often triggered by or associated with ethnobotanical background, health claims and beneficial properties of new ingredients (e.g. Atala, 2012 and Renna and Gonnella, 2012). Furthermore, flavouring of foods using natural herbs and spices is a preferred approach to reduce salt use in dish preparation and to increase consumer acceptability and appreciation (Ghawi et al., 2014).

Dittany of Crete (*Origanum dictamnus* L.) is a unique aromatic cushion-forming perennial plant of Lamiaceae family (Mint family) which is a local endemic to the Island of Crete, Greece (Fig. 1). In the wild it grows exclusively as a rock-dweller in crevices and rocky habitats, generally from 300 m or more commonly from 500 m to 1800 m above sea level (Turland, 1995). Its above ground parts (non-woody parts) have been traditionally used in Greece as a comminuted herbal substance for preparation of an infusion (locally called 'vrastari' in Crete, Greece, from the verb *vrazo*=to boil) or tincture by decoction for cutaneous use; in its native range, it is occasionally chewed crude against gingivitis cough and cold (Skoula and Kamenopoulos, 1997). Concerning its culinary use, inflorescences and leaves have been added sometimes to sauces, salads and vermouth. Dried leaves and extracts have also been reportedly used in bitters, liqueurs and fish sauces (The Herb Society of America, 2005 and Chishti et al., 2013). However, its culinary use is limited (The Herb Society of America, 2005 and Chishti et al., 2013).

Figure. 1. Dittany of Crete (*Origanum dictamnus* L.) in the wild, growing exclusively as a rock-dweller in crevices and rocky habitats (Photo: E. Kalpoutzakis).

In our study we review current literature for the Dittany of Crete (*O. dictamnus*) from an ethnobotanical point of view and we explore its potential for culinary use in the Mediterranean cuisine, in an attempt to exploit the traditionally known beneficial effects of this plant (EMA/HMPC, 2013). Our study expands the use of a unique and beneficial herb (*O. dictamnus*) rendering it as a new spicy ingredient suitable for gastronomic experimentation and study.

2. MATERIAL AND METHODS

We explored the vernacular names of the Dittany of Crete in the Greek language in an attempt to document its long-standing traditional use and to explore its long-standing traditional use in respect to safety rules for human consumption as a food.

We conducted extensive searches in the Scopus and Google Scholar bibliographic databases to find previously published studies indicating uses for the Dittany of Crete. We consulted the scientific literature to find the medicinal properties which are attributed to the Dittany of Crete and to document whether consolidated indications exist according to the requirements of the European Medicines Agency (http://www.ema.europa.eu/ema/).

We reviewed literature sources (Phitos et al., 1995, Phitos et al., 2009 and Bilz et al., 2011) to define whether there is some protection status designated for the exploitation of wild populations and if an extinction risk is assigned to the wild populations of this local endemic plant of Crete (Greece). We also reviewed the history of its commercial cultivation in Greece.

By testing several ingredients, we explored the possibility of cooking different dishes (appetisers, salads, main dishes, desserts) with Ditanny of Crete, using fresh or dried material in different quantities and their infusions; this procedure resulted in the culinary concept presented in this study.

To assess initially the quality, odour, taste and the overall sensory properties of the case-studies developed using different quantities of plant material, we formulated a non-official tasting panel. This first round sensory evaluation panel included a local chef, a food-editor, two everyday cooks and two adults with no previous gustative training (in total, 2 men and 4 women). The non-formal assessors were asked to taste freely from the samples presented (as much as or as often as they wanted) and to provide us with their comments during a long discussion with the authors. The first assessment included tasting of crude and chopped plant material, the second round included samples of infused plant material (for 10 persons, 30 g of plant material infused in 1.5 L of hot water for 4–5 min), while the third, fourth and fifth rounds were targeted to the prepared dishes using different quantities of Dittany of Crete during preparation (case-studies developed). All samples (crude plant material, infused plant material and prepared dishes), were presented to the panellists at room temperature under normal lighting conditions and the evaluation session was held in the foyer of the Laboratory of Conservation and Evaluation of Native and Floricultural Species (Thessaloniki, Greece). The case-studies developed were prepared before the session and were served to the panellists in the afternoon of the same day.

3. RESULTS AND DISCUSSION

3.1 Vernacular names suggesting ancient use

There is strong ancient mythological and historical context about the Dittany of Crete, resulting to several vernacular Greek names which reflect its long-standing traditional use by humans. In total, we detected at least 10 vernacular names related to the Dittany of Crete.

Archaebotanical evidence (seeds) of *O. dictamnus* were found during excavations in the Knossos palace near Heraklion, Crete; these date back to Minoan times (27th–15th century B.C.), suggesting a really ancient use of this plant (Diapoulis, 1949).

Originally, the plant was dedicated to the ancient Cretan goddess Diktynna, thus elucidating its ancient common name Dittany (*Dictamos* in Greek). However, *Dictamos*may also derive from *Dicti*, the name of the Cretan mountain where Zeus (Jupiter) was raised up by the goat

Amalthia and the Greek word *thamnos*=shrub (Skrubis, 1979 and Skoula and Kamenopoulos, 1997).

The power of this plant has been praised by numerous writers from antiquity up to the fourth century A.D. (EMA/HMPC, 2013). Hippocrates (5th–4th century B.C.) used it to cure gastric complaints, tuberculosis and in poultices on wounds. Aristotle (Historia Animalia 9.16.1), and later Theophrastus (4th–3rd century B.C.) (Thanos, 1994), Dioscorides (1st century A.D.) (Berendes, 1902), Galen and Plutarch (1st–2nd century A.D.) (Liolios et al., 2010), all state that wild goats of Mount Ida (Crete) when struck by poisoned arrows used to eat shoots of *O. dictamnus* to heal their wounds. This tradition passed to Romans and famous Latin writers such as Cicero (2nd–1st century B.C.) and Virgil (1st century B.C.) attributed the healing of the wounds of the Trojan hero Aeneas to the Dittany of Crete (Virgil, Aeneid 12.412) (Hunt, 2005). Some ancient vernacular names of Cretan Dittany like *veloulko* or *velotoko* (velos=arrow) probably elucidate the plant's ability to heal arrow wounds (Gennadios, 1914). Other vernacular Greek names include stomachohorto (from Greek *stomachos*=stomach, referring to its healing properties to cure stomach ache, and *horto*=herb, Plimakis, 1997 and Skoula and Kamenopoulos, 1997), kephalohorto (from Greek *kephali*=head), stamatochorto [literally meaning in Greek 'stopping something herb', originating from the Greek verb *stamato* (= to stop)], livanohorto (from Greek *livani*=incent), maliarohorto (from Greek*maliaros*=hairy), and *gerondas* (=old man); the latter two, evidently refer to the plant's morphology, as its leaves are covered by dense white hairs (Skoula and Kamenopoulos, 1997 and Liolios et al., 2010; see also Fig. 1).

Due to difficulties in reaching the plant's wild habitats, it was traditionally believed that someone has to be truly in love in order to climb and gather the Dittany shoots to be offered (Plimakis, 1997 and Fragaki, 1969); this reflects another vernacular name i.e. Erontas (from the ancient Greek God *Eros*=love).

According to EMA/HMPC (2013), this plant has also common names in English (Dittany of Crete, Dictamna hops, Dictamnus, Winter Sweet, Dittany, Hop Marjoram, Hop Plant), Turkish (Mangirotu), German (Diptamdosten, Kretadiptam, Dictamno, Kreta-Majoran), Italian

(Origano di Creta, Dittamo di Candia, Dittamo Cretico) and French (Dictame de Crête).

3.2 Dittany of Crete: a traditional herbal medicinal product and a possible functional food

Dittany of Crete is a famous medicinal plant since ancient times. It was considered as 'panacea' (universal remedy) and it was thoroughly used for stomach disorders, gastric ulcers, generally for the peptic system, spleen, against rheumatism, to facilitate childbirth and against gynaecological disorders (EMA/HMPC, 2013). It was also widely used in medieval monasteries, in the famous liqueurs Benedictine and Trappistine of the Benedictines and Trappistines monks, respectively (EMA/HMPC, 2013). Even in our days, it is used in distilleries, as vermouth for example is flavoured with extracts of this aromatic plant (Baumann, 1996 and Skrubis, 1979). During the middle ages, the plant is recorded to the code of Carlomagnus (742–814 A.D.) (EMA/HMPC, 2013).

The non-woody above ground parts of the plant contain essential oils rich in volatile compounds (1.5–2.5%) such as carvacrol, γ-terpinene, thymol and p-cymene (Sivropoulou et al., 1996, Skoula et al., 1999, Daferera et al., 2000, Chorianopoulos et al., 2004 and Liolios et al., 2010) (Fig. 2). Dittany of Crete also contains high concentration of polyphenolic compounds, mainly depsides (Fig. 3) (Chatzopoulou et al., 2010, Exarchou et al., 2013, Kaliora et al., 2014, Lagouri and Alexandri, 2013, Skotti et al., 2014 and Tair et al., 2014), coumarins, triterpenes (e.g. the rare 21-aOH oleanolic acid), tocopherols (e.g. Lagouri and Boskou, 1996), sterols (e.g. Komaitis et al., 1988) and fatty acids (e.g. palmitic, oleic, linoleic acids, Revinthi-Moraiti et al., 1985).

Dittany of Crete has been associated with several medicinal properties: it is orexigenic (triggers appetite), antioxidant (e.g. Couladis et al., 2003, Kaliora et al., 2014, Kouri et al., 2007, Lagouri and Alexandri, 2013, Lionis et al., 1998, Møller et al., 1999, Skotti et al., 2014 and Tair et al., 2014), diuretic and digestive (e.g. Bazaios, 1982), emmenagogue (e.g. Bazaios, 1982), antispasmodic (e.g. Skoula and Kamenopoulos, 1997), astringent, soothing the nervous system (e.g. Bazaios, 1986). Additionally, it has antibacterial (Sivropoulou et al., 1996 and Economakis et al., 2002), antifungal (Vokou et al., 1993,Daferera et al., 2000 and Karanika et al., 2001) and anthelmintic properties (e.g. Skoula and Kamenopoulos, 1997).

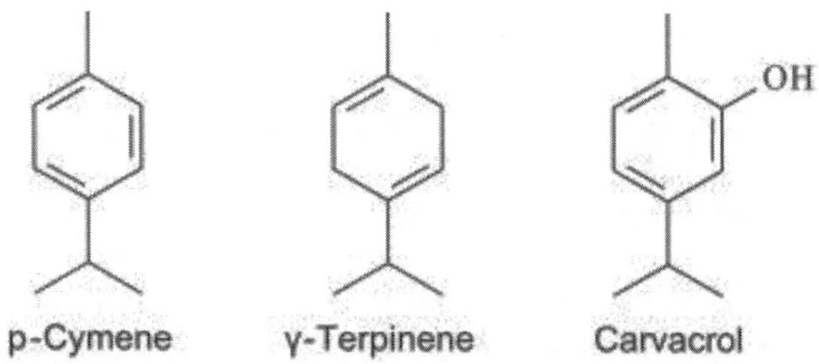

Figure. 2. Structures of the major volatile compounds of essential oils of *Origanum dictamnus* (Lamiaceae).

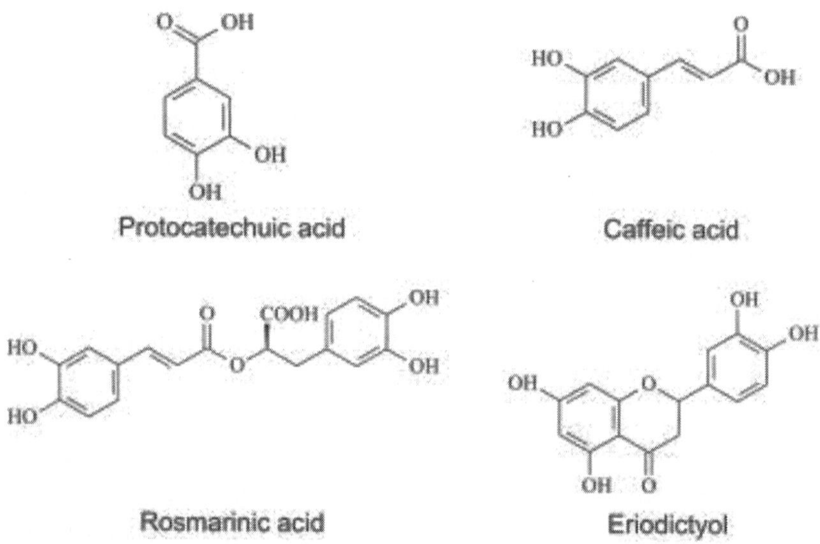

Figure. 3. Major polyphenolic compounds found in the polar extracts of *Origanum dictamnus* (Lamiaceae).

There is an evidence on traditional medicinal use in Greece for 'at least 30 years' (literature references in accordance for 40 years, actual use for >50 years), as requested by EU Directive 2004/24/EC and its uses are confirmed by a large number of publications providing consistent information (EMA/HMPC, 2013 and references therein). In this respect this plant is considered safe for consumption as a spice. Dittany of Crete and its infusion preparation is regarded to date as traditional herbal medicinal product in the following indications (EMA/HMPC, 2013): (i) for the relief of cough associated with cold, (ii) for the relief of minor skin inflammations and bruises and (iii) for the relief of mild gastrointestinal disorders. The latter indication could be exploited in gastronomy; cooking ingredients prepared with Dittany of Crete may prove to offer additional functions to food, thus rendering Dittany as a possible fortifying functional element to be studied further.

3.3 Expanding current uses of Dittany of Crete

The main use of Dittany of Crete is currently restricted to the preparation of herbal teas (infusion, decoction) for oral use, tinctures or for cutaneous use (Chishti et al., 2013,Liolios et al., 2010 and EMA/HMPC, 2013).

According to current knowledge, the culinary use of Dittany of Crete is too limited or not recommended. None of *Origanum* experts recommend this species for culinary use, claiming that the scent is too medicinal (therefore not attractive and savoury) and the flavor is too strong for cooking (The Herb Society of America, 2005 and Chishti et al., 2013). Although Dittany of Crete is considered as orexigenic, antioxidant, digestive, with considerable antimicrobial properties, and despite the fact that it is assessed as a safe and effective traditional herbal medicinal product especially for gastrointestinal disorders (EMA/HMPC, 2013 and references therein), it has not actually been used as a spice in dish preparations. The densely hairy leafs of Dittany, its medicinal scent and strong flavour (The Herb Society of America, 2005 and Chishti et al., 2013) have probably restricted its culinary use to date. However, when used in certain quantities, its culinary use becomes challenging and attractive (see case-studies developed below).

3.4 Threatened wild growing populations and commercial interest

To date Dittany of Crete is a plant with significant commercial interest since it is widely used for herbal teas in Crete and mainland Greece. In the past, the demand for plant material was covered exclusively by harvesting wild populations. Local people associated with collections of Dittany of Crete from the wild are called in Greek 'Erontades', 'Atitanologi' or 'Mazoctades', indicating harvesting pressure since ancient times. This probably led to rapid decrease of wild populations and some local extinction events from several areas on the island of Crete where it is endemic.

Nationally it has been designated as threatened with extinction (Vulnerable, according toTurland, 1995) and it was recently assessed at the European scale as Near Threatened (Bilz et al., 2011). Dittany of Crete is considered as a protected plant due to its inclusion in the Presidential Decree 67/1981 at the national level (this regulates any access to its wild populations). It is also covered by the Bern Convention at the international level. In addition, *O. dictamnus* is included as a Priority Species in Annexes II and IV of the European Directive 92/43/EEC (Habitats Directive); for this plant, EU (European Union) member states (for Dittany i.e. Greece) are entirely responsible for its long-term conservation globally.

The plant's relative rarity, the climbing accidents occurred during quests for the plant and the increasing commercial demand for plant material, probably induced its cultivation since 1920s (in Embaros, South Crete). After 1935, there was a rapid increase in its cultivation in other places of Crete reaching a peak production of 10 t per year. Right after 1945, the production reached 50 tonnes per year and to date almost 85% of the product is exported (mainly to Italy, France, Germany and Japan), while 15% of the total production is absorbed by the Greek market (Skoula and Harborne, 2002). The plant material can be purchased in traditional or modern shops with local products and open-air market stalls (Hanlidou et al., 2004, Liolios et al., 2010 and Skoula and Kamenopoulos, 1997). Although the plant is known in several countries, no marketed products seem to be available outside Greece (EMA/HMPC, 2013).

Dittany of Crete as a new ingredient in gastronomy: culinary concept, reasoning, process and quantities

The aim of the proposed culinary concept was the integration of the flavours of the Dittany of Crete into the Mediterranean cuisine and the exploitation of its beneficial properties in dish preparation; this represents a new trend in international gastronomy and food science (Atala, 2012). Our reasoning was triggered by the fact that Dittany's oral use as herbal tea is widespread in Greece with well-established indications (EMA/HMPC, 2013). Hence, we reasoned that if the plant's aromatic infusion is so much praised for well-being (especially for the peptic system), to incorporate its flavour into food preparation should be equally savoury and beneficial. In this way, we explored possible uses of Dittany of Crete in dish preparation.

The inflorescences and the leaves (fresh or dried) of Dittany of Crete and their infusion (Fig. 1 and Fig. 4) can be a novel and an interesting ingredient in cooking which can be exploited in a wide range of different dishes, offering additionally significant medicinal properties (Fig. 5, Fig. 6 and Fig. 7). Inflorescences and leaves can be consumed crude and a traditional simple technique i.e. infusion can be used to collect this new flavour, thus providing new perceptions in gastronomy (Fig. 5, Fig. 6 and Fig. 7). The basic flavour of Dittany of Crete is strong, sharp and slightly bitter balancing between oregano-like (mainly due to carvacrol, Fig. 2) or thyme-like (due to thymol, Liolios et al., 2010) nuance.

3.5 Cuisine applications

We developed the following case-studies combining infusion, leaves and/or inflorescences of Dittany of Crete (Fig. 5, Fig. 6 and Fig. 7):

- *Cherry pie with dark chocolate and syrup from Dittany of Crete* [dough: cereal flour, water, egg, yeast, chopped Dittany leaves, brown sugar, olive oil, ginger, salt; filling: dark chocolate, cherries; syrup: brown sugar, one cup with infusion of Dittany, Fig. 5].

- *Beef meatballs with marinated oven leeks and béchamel sauce with Dittany of Crete*[meatballs: no-crust bread, beef mince, milk, salt, pepper; leeks: marinated with salt, olive oil, red wine and chopped Dittany leaves and inflorescences, then baked in the oven; béchamel: olive oil, cereal flour, milk, one cup of Dittany infusion, eggs, grated hard goat and sheep's cheese, salt and black pepper, Fig. 5].

- *Fish-egg salad with Dittany of Crete* [leavened bread, fish roe 'white tarama', olive oil, one cup with Dittany infusion, chopped fresh Dittany leaves and inflorescences, lemon juice, Fig. 6].

- *Marinated rabbit with pearl onions and sweet potato* [marinade: red wine, laurel, one cup with infusion of Dittany of Crete; cooked in Dutch oven, Fig. 6].

- *Fresh salad of climbing French beans boiled with Dittany of Crete, with Greek yogurt sauce, topped with Sun flower seeds* [climbing French beans boiled with chopped Dittany leaves and inflorescences; cow yogurt with olive oil, vinegar, lemon juice, salt, black pepper, Fig. 7].

Figure. 4. Fresh shoots of Dittany of Crete (local Greek endemic plant) collected in early summer from sustainable cultivation of propagated plants in the Balkan Botanic Garden of Kroussia (Greece) and colour of its traditionally made infusion (3 g of plant material in 150 ml of hot water, after 4–5 min).

Figure. 5. Case-studies of cuisine applications prepared with *Origanum dictamnus* infusion, leaves and/or inflorescences: Left: Cherry pie with dark chocolate and syrup from Dittany of Crete; the aromatic effect of Dittany is hidden in the dough and the sweet syrup. Right: Beef meatballs with marinated oven leeks and béchamel sauce with Dittany of Crete; in this complex case with many ingredients, the Dittany marinates the leeks in the oven and offers original aroma to the béchamel sauce.

Figure. 6. Case-studies of cuisine applications prepared with *Origanum dictamnus* infusion, leaves and/or inflorescences: Left: Fish egg salad with Dittany of Crete; in this case the Dittany's scent is perfectly blended with the flavours of the rest of the ingredients. Right: Marinated rabbit with pearl onions and sweet potato; in this case, the rabbit is marinated with wine, Dittany and laurel overnight and the oregano-like or thyme-like savoury effect is contrasted with the pungent and bland tastes of onions and sweet potatos, respectively.

Figure. 7. Fresh salad of climbing French beans boiled with Dittany of Crete, served with Greek yogurt sauce and topped with Sun flower seeds: a case-study prepared with *Origanum dictamnus* infusion of leaves and/or inflorescences; boiled Dittany leaves and/or inflorescences for 15 min provide the intriguing aftertaste (the floral decorative of inflorescence bracts may also be eaten crude for stronger flavour).

Given that the scent of Dittany of Crete has been considered as too medicinal or too strong for cooking (The Herb Society of America, 2005 and Chishti et al., 2013), for the case-studies developed (see Fig. 5, Fig. 6 and Fig. 7) we first used rather small quantities of Dittany i.e. for 4 persons 2 g of plant material and 3 g infused in 150 ml of hot water for 4 min; these attempts resulted in interesting and original mild flavours according to the first round assessors. Although flavour in some dishes may be complex due to several ingredients used (e.g. dishes in Fig. 5), the aromatic intensity of Dittany is still distinctive in the quantities mentioned above. We then doubled and tripled the quantities used for the case-studies in pursuit for stronger flavours. When the quantity used was doubled, the interaction of flavours delivered an intense mouth feel sensation with sudden fresh explosions of aromas by the chewing of chopped Dittany, which easily travels up from mouth to nose according to the opinion of the first round assessors. When the quantity used was tripled, the flavours became rather intense and aggressive and the sensorial perception obtained a strong and lingering aftertaste; some of the panellists tasting these samples found them too strong for their tastes.

All the case-applications developed in this study were prepared using smaller quantities of Dittany of Crete than the single daily doses (4.5–21 g) proposed by EMA/HMPC (2013). Although no overdose case or drug abuse has been recovered in the scientific literature, it is important to note that the safety of Dittany of Crete for children and adolescents or during pregnancy and lactation has not been established yet (no specific studies exist in literature). EMA/HMPC (2013), in accordance with general medical practice, suggests that "the herbal medicinal products (herbal teas and other finished products) should not be used during pregnancy and lactation without medical advice".

4. CONCLUSIONS

The culinary concept presented in this study provides new sensations to different food products which may contribute at the same time to human well-being. Second, but not less important, a rational and sustainable exploitation of this plant in a wider scale may be promoted through gastronomy, offering important socio-economic benefits to the local population of Crete. This has recently been also attempted for other ingredients originating from biodiversity-rich areas such as Amazonia, Brazil (Atala, 2012). If Cretan Dittany is successfully promoted as a food ingredient (or a possible functional food) to be launched in easily made dishes of everyday life and in *nouvelle cuisine* preparations, it could soon be considered as a high value-added product which can offer potential to the sustainable development at local scale. This could be quite significant especially in countries severely affected by the deep economic crisis (such as Greece). Being a local endemic of Crete (Greece), if the market demand is increased for this food product, any future commercialisation by international companies will have to adapt to the provisions of the Nagoya Protocol and comply with national rights on access to endemic genetic resources, involving special benefit and sharing mechanisms (Kamau et al., 2010).

To date, the natural products derived from plants show a great appeal for consumers and chefs and as a consequence the introduction of new spices from plants could stimulate new opportunities for gastronomy (Renna and Gonnella, 2012). Dittany of Crete meets all necessary requirements to be introduced as a new, interesting and unique spice for culinary use which is accompanied with important medicinal properties, thus rendering it as a potential functional food. The promotion of new uses for this traditionally used medicinal plant (currently cultivated only at small scale on the island of Crete) (i) offers new ingredients to international gastronomy, (ii) may prove to be beneficial for local economies, and (iii) supports sustainable plant exploitation.

REFERENCES

1. Atala, A., 2012. Ane wing redient: the introduction of priprioca ingastronomy. Int. J.Gastron.FoodSci.1(1),61–63.

2. Baumann, H., 1996. Greek wild Flowers and Plant Lore in Ancient Greece. The Herbert Press Ltd., London.

3. Bazaios, K., 1982. 100 Herbs for 1000 therapies. Athens.

4. Berendes, J.,1902. Des Pedanios Dioskurides aus Anazarbos Arzeinmittellhre in fünf Büchern availableat:. Verlagvon Enke, Stuttgart 284–285.

5. Bilz, M., Kell, S.P., Maxted, N., Lansdown, V.R., 2011. European RedListof Vascular Plants. Publications Office of the European Union, Luxembourg.

6. Chatzopoulou, A., Karioti, A., Gousiadou, Ch., Lax Vivancos,V., Kyriazo- poulos, P., Golegou, S.,Skaltsa, H.,2010. Depsides and other polar constituents from Origanum dictamnus L. and their invitro antimicrobial activity inclinicalstrains. J. Agric. FoodChem. 58,6064–6068.

7. Chishti, S.,Kaloo,Z.A.,Sultan,P.,2013.Medicinalimportanceofgenus Origanum: areview. J.Pharmacogn.Phytother.5(10),170–177.

8. Chorianopoulos,N.,Kalpoutzakis,E.,Aligiannis,N.,Mitaku,S.,Nychas,G.-J., Haroutounian, S.A.,2004.Essentialoilsof Satureja, Origanum and Thymus species:chemical composition and antibacterial activities against food borne pathogens.J.Agric.FoodChem.52,8261–8267.

9. Couladis, M.,Tzakou,O.,Verykokidou,E.,Harvala,C.,2003.Screeningof some Greekaromatic plants for antioxidant activity. Phytother. Res.17(2), 194–195.

10. Daferera, D.J., Ziogas, B.N., Polissiou, M.G.,2000.GC–MS analysisof essential oils from some Greek aromatic plants and their fungi toxicityon Penicillium digitatum. J.Agric. Food Chem.48(6),2576–2581.

11. Diapoulis, H.A.,1949. Eliniki Chloris (Synopsis Florae Graecae),vol.2. Minestère del'Agriculture, Athens [inGreek].

12. Economakis, C.D., Skaltsa, H.,Demetzos, C.,Sokovic, M.,Thanos, C.A., 2002. Effect of phosphorus concentration of the nutrient solution on the volatile constituents of leaves and bracts of Origanum dictamnus. J. Agric. Food Chem.50(22),6276–6280.

13. EMA/HMPC, 2013. Final assessment report on Origanum dictamnus L., herba (200431/2012). European Medicines Agency/ Committee on Herbal Med- icinal Products, London (availableat: ⟨http:// www.ema.europa.eu/docs/ en_GB/document_library/Herbal_-_HMPC_assessment_report/2013/08/ WC500147179.pdf⟩).

14. Exarchou,V.,Takis,P.G.,Malouta,M.,Vervoort,J.,Karali,E.,Troganis,A.N., 2013.Fournewdepsidesin Origanumdictamnus methanolextract. Phytochem. Lett.6,46–52.

15. Fragaki, E.K., 1969. Contribution to the plant local names. Athens [in Greek]. Gennadios, P. 1914. Lexicon Phytologicon. Athens [in Greek].

16. Ghawi, S.K.,Rowland,I.,Methven,L.,2014.Enhancing consumer liking of low salt tomatos oupover repeated exposure by herband spice seasonings. Apetite 81(1),20–29.

17. Hanlidou, E.,Karousou,R.,Kleftogianni,V.,Kokkini,S.,2004.Theherbal market ofThessaloniki(N-Greece)anditsrelationtotheethnobotanical tradition. J.Ethnopharmacol.91,281–299.

18. Hunt, P.,2005.AeneidXII.383-440asinspirationforancientart: the Roman surgeon (availableat: 〈http:// traumwerk. stanford.edu/ philolog/2005/11/ aeneid_as_inspiration_for_anci.html〉).

19. Kaliora, A.C.,Kogiannou,D.A.A.,Kefalas,P.,Papassideri,I.S.,Kalogero-poulos, N.,2014.Phenolicprofiles andantioxidant and anticarcinogenic activities of Greek herbal in fusions; balancing delight andchemopreven- tion. FoodChem.142,233–241.

20. Kamau,E.C.,Fedder,B.,Winter,G.,2010.TheNagoyaProtocolonaccessto genetic resourcesandbenefit sharing: what is new and what are theimplications for provideranduser countriesandthescientific community? LawEnviron. Dev. J.6(3),1–246 (availableat〈http:// www.lead-journal.org/content/ 10246.pdf〉.

21. Karanika, M.S.,Komaitis, M.,Aggelis, G.,2001. Effect of aqueous extracts of some plantsofLamiaceaefamilyonthegrowthof Yarrowia lipolytica. Int. J. FoodMicrobiol.64(1-2),175–181.

22. Komaitis, M.E.,Revinthi-Moraiti,K.,Evangelatos,G.,1988.Thelipid composition of fresh Origanum dictamnus leaves. FoodChem.27(1), 25–32.

23. Kouri, G.,Tsimogiannis, D.,Bardouki, H.,Oreopoulou, V.,2007. Extraction and analysisofantioxidantcomponentsfrom Origanum dictamnus. Innov. Food Sci.Emerg.Technol.8,155–162.

24. Lagouri, V.,Boskou, D.,1996. Nutrient antioxidant sinoregano. Int.J.Food Sci. Nutr.47(6),493–497.

25. Lagouri, V.,Alexandri,G.,2013.AntioxidantpropertiesofGreek O. dictamnus and R. officinalis methanol andaqueousextracts—HPLC determination of phenolicacids.Int.J.FoodProp.16,549–562.

26. Liolios, C.C.,Graikou,K.,Skaltsa,E.,Chinou,I.,2010.DittanyofCrete:a botanical andethnopharmacologicalreview.J.Ethnopharmacol.131, 229–241.

27. Lionis, C.,Faresjo,A.,Skoula,M.,Kapsokefalou,M.,Faresjo,T.,1998. Antioxidant effects of herbs in Crete.Lancet (NAmericanEd.) 352(9145), 1987–1988.

28. Morris, J., 2012. Super-Food Kitchen: Cooking With Nature's Most Amazing Foods. Sterling Epicure, New York.

29. Møller, J.K.S.,Madsen,H.L.,Aaltonen,T.,Skibsted,L.H.,1999.Dittany (Origanum dictamnus) asasourceofwater-extractable antioxidants. Food Chem. 64(2),215–219.

30. Phitos, D., Strid, A., Snogerup, S., Greuter, W. (Eds.), 1995. The Red Data Book. World Wide Fund for Nature (WWF), Athens, Greece.

31. [in Greek].Phitos,D.,Constantinidis,T.,Kamari,G.(Eds.),2009.TheRed Data BookoftheRareandThreatenedPlantsofGreece,vol.1.Hellenic Botanical Society,Patras(A–D) andvol.2(E–Z).

32. Plimakis, A.G., 1997. Dictamnus the Miraculous Herb of Crete. Hania, Greece [in Greek].

33. Renna, M.,Gonnella,M.,2012.Theuseoftheseafennelasanewspice-colorant inculinarypreparations.Int.J.Gastron.FoodSci.1(1),111–115.

34. Revinthi-Moraiti,K.,Komaitis,M.E.,Evangelatos,G.,Kapoulas,V.,1985. Identification and quantitative determinationofthelipidsofdried Origanum dictamnus leaves. FoodChem.16,15–24.

35. Sivropoulou,A.,Papanikolaou,E.,Nikolao,C.,Kokkini,S.,Lanaras,T., Arsenakis, M.,1996.Antimicrobialandcytotoxicactivitiesof Origanum essential oils.J.Agric.FoodChem.44(5),1202–1205.

36. Skotti, E.,Anastasaki,E.,Kanellou,G.,Polissiou,M.,Tarantilis,P.A.,2014. Total phenolic content, antioxidant activity and toxicity of aqueous extracts from selected Greek medicinal an daromatic plants.Ind.Crop.Prod.53, 46–54.

37. Skoula, M.,Gotsiou, P.,Naxakis, G.,Johnson, B.C.,1999. Achemosystematic investigation onthemono-and sesquiter penoids in the genus Origanum (Labiatae). Phytochemistry52(4),649–657.

38. Skoula, M.,Harborne,J.B.,2002.Oregano:thegenera Origanum and Lippia. In: Kintzios,S.E.(Ed.),TaxonomyandChemistryof Origanum. Medicinal and AromaticPlantsIndustrialProfiles. CRC Press, Taylon& Francis, London, pp.67–108.

39. Skoula, M.,Kamenopoulos,S.,1997. Origanum dictamnus L. and Origanum vulgare L. subsp. hirtum: traditional uses and production in Greece. In: Padulosi, S.(Ed.),. Proceedings of the IPGR IInternationalWorkshopon Oregano, 8–12 May 1996, CIHEAM, Valenzano,Bari,Italy,pp.26–32.

40. Skrubis, B.,1979. Origanum dictamnus L., aGreeknativeplant.J. Ethnopharmacol.1(4),411–415.

41. Tair, A.,Weiss,E.-K.,Palade,L.M.,Loupassaki,S.,Makris,D.P.,Ioannou,E., Roussis, V.,Kefalas,P.,2014. Origanum species nativetotheIslandof Crete: invitroantioxidantcharacteristicsandliquidchromatography – mass spectroscopyidentification of major polyphenolic components. Nat. Prod. Res.28(16),1284–1287.

42. Thanos, C.A.,1994.AristotleandTheophrastusonplant-animal interaction. In: Arianoutsou,M.,Groves,R.H.(Eds.),Plant-Animal Interactions in Mediterranean-Type Ecosystems. Kluwer Academic Publishers,Dordrecht, The Netherlands,pp.3–11.

43. The Herb Society of America. 2005. Oregano and Marjoram, an Herb Society of America guide to the genus Origanum. , Kirtland, Ohio,USA.

44. Turland, N.J.,1995. Origanum dictamnus L., Vulnerable (VU). In: Phitos,D., Strid, A., Snogerup, S., Greuter, W.(Eds.), The Red Data Book. World Wide FundforNature(WWF),Athens,Greece,pp.394–395.

45. Vokou, D.,Vareltzidou, S.,Katinakis, P.,1993. Effects of aromatic plants on potato storage: sprout suppression and antimicrobial activity. Agric. Ecosyst. Environ.47,223–235.

12

USES OF RHIZOPUS ORYZAE IN THE KITCHEN

Igor Cantabrana[a,,], Ramón Perise[a], Igor Hernández[b]

[a] Mugaritz, Otzazulueta Baserria, Aldura Kalea 20 zk, Errenteria 20100, Gipuzkoa, Spain

[b] Fundación AZTI-AZTI-Fundazioa, Food Research Divison, Parque Tecnológico de Bizkaia, Astondo Bidea, Edf. 609, 48160 Derio, Bizkaia, Spain

ABSTRACT

The use of fungal cultures in modern cuisine can provide a broad number of new textures, flavors and tastes from unexpected substrates; such as fruits, vegetables and nuts. The presented research describes how *Rhyzopus oryzae*, a fungi used in Asian culture to produce traditional recipes, results in fruits, vegetables and grains with unique sensorial properties. Throughout the paper, different examples of these novelty uses are presented showing different examples of prototypes those have been successfully incorporated into our menus, their production procedures and their sensorial evaluations.

KEYWORDS

Rhyzopus oryzae; Mold fermentations; Innovative gastronomy

1. INTRODUCTION

Tempeh is a popular fermented product in Indonesia and Malaysia that includes, traditionally, fresh or cooked soybean and a mixture of fungus, yeast and bacteria that ferments the wet seed and produces a solid paste that can be fried, boiled or consumed raw (Shurtleff and Aoyagi, 2001). During fermentation, microorganisms use the seed as a substrate to feed themselves and subsequently obtain energy and organic material. It is a complex process where fungal metabolism plays a key role and results in significative changes in the texture, flavor and taste of the vegetal material. Although tempeh is not usually commercialized in Western countries, it is widely used in Asia and it has become an emergent food industry. Transformation from artisanal to modern food production requires pure, standardized and well-defined starter strains. They make it possible to expand the uses of these microorganisms in different food products and can yield more consistent results in controlled conditions and at production facilities.

In most cases, traditional products are fermented with a mixture of yeast, bacteria and fungi. Among others genus, *Rhizopus* is part of the fungal ecosystem responsible of these fermented products. Scientific classification of this fungus is Class, *Phycomycetes*; Order, *Mucorales*; Family, *Mucoraceae*; Genus, *Rhizopus.* This genus is composed of 10 species, including plant spoilage species (like *Rhizopus arrhizus* or *Rhizopus artocarti*) or food related species, like *Rhizopus oligosporus* and *Rhizopus oryzae*, those involved in tempeh production (Wiesel et al., 1997). Different characteristics make this genus interesting for massive application in food production. *Rhizopus* requires a simple ecosystem to survive and can grow vigorously between 25 and 45 °C. All these characteristics make them almost omnipresent in nature, and allow almost all vegetables material colonization. The main constraint to fungal growth is humidity, which should remain at an elevated percentage. During fermentation, *Rhizopus'* amylase, lipase and protease activity (Baumann and Bisping, 1995), increases the biodisponibility of the nutrients and their ability to use many compounds as energy and carbon source. Also, this fungus has a very rich secondary metabolism, it produces a high number of compounds with sensorial and nutritional interest (Denter et al., 1998).Finally, *Rhyzopus*can produce anti-fungal and anti-bacterial compounds, which have been reported byDinesh Babu et al. (2009).

As with other microorganisms, safety concern is critical when human food application is considered. *R. oryzae* is considered GRAS (Generally Recognized As Safe) by the U.S Food and Drug Administration (FDA) and thus, can be used for human consumption within the U.S. However, filamentous fungi is not considered QPS (Qualified Presumption of Safety) by the EFSA (European Food Safety Agency), which raise safety concerns, in which mycotoxins presence therefore should be analyzed in individual products marketed in Europe. To our best of knowledge, no toxin production by *R. oryzae*has been reported in scientific literature. Actually, the *Rhizopus* species has been used on the one hand, as a detoxifying agents against food toxins, like Ochratoxin A (Varga et al., 2005), and on the other hand, to increase the digestibility of certain legumes (Azeke et al., 2007).

One of the main advantages of working with this microorganism in the kitchen is that molds open a window in sensorial world in terms of new textures and flavors. In the case of *R. oryzae*, there is a special interest to obtain innovative results in matrix apart of traditional soy beans. Among of this, molds change nutritional characteristics during fermentation, those may improve nutritional profile in menus. Examples of this modifications are described in Baumann and Bisping (1995), in Denter et al. (1998) and in Dinesh Babu et al. (2009).

In Western gastronomy, mold-fermentations are not usual, except significant exceptions (i.e. Roquefort-like cheeses). In this paper we present examples of different products that can be obtained using *R. oryzae* as a fermentation agent, in an effort to communicate the amazing textures and flavor modifications that can be induced by fungal fermentation from an exclusively gastronomic point of view. This paper does not try to be an exhaustive relation of fermentable substrates and we encourage each cook to test with their own "magical recipe".

2. MATERIALS AND METHODS

2.1 Used microorganisms, ingredients and cooking conditions

Commercial *R. oryzae* culture was provided by Top Cultures, (Belgium). The inoculum was stored at 4 °C. The specification sheet claims 8,000,000 spores per gram of starter when it was packaged. Starter was inoculated without previous rehydratation. In all the cases described in the paper, raw material was cooked in tap water and 200 g of cooked substrate was inoculated with 1 g of dry starter (see Fig. 1). Preliminary results (not shown) demonstrated that using smaller ratio of inoculum produce incomplete colonization and, visually, very heterogeneous results.

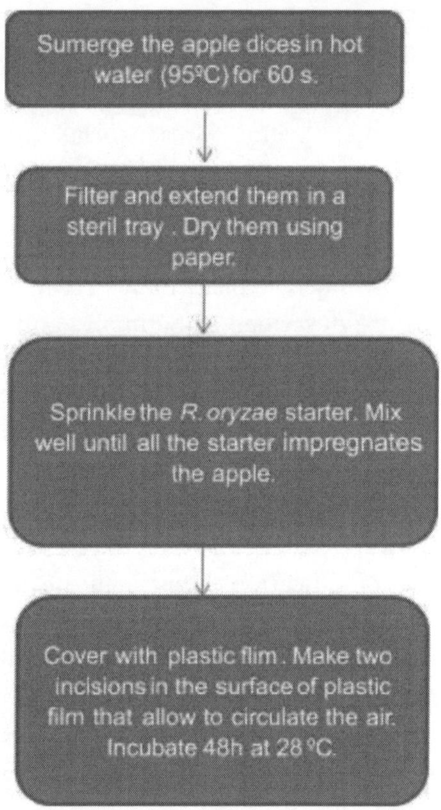

Figure. 1. Workflow diagram for apple-dice fermentation.

Used raw materials (Golden and Granny Smith apples, rice, seeds cereals and legumes) were provided by local markets. Cooking was done using habitual material present in professional installations, including induction cooking facilities and industrial ovens with air circulation.

2.2 Inoculation and growth; safety considerations

Fungal growth requires long incubation periods (24–48 h) at 28 °C, which would make the food matrix a good substrate for possible pathogens. Incubation conditions as well as the absence of later thermal treatment of most of the products make food safety our first concern. Possible contamination with pathogens should take in mind and extremely hygienic practices should be observed, including some modification in the usual kitchen practices. Among others, all material in contact with the ingredients, including knives, bowls and trays should be cleaned with 70% ethanol before use. Gloves should be used and cleaned frequently with 70% ethanol. To inoculate efficiently, one recommends use tools that allow the starter to spread on different foods. In this case, the starter is a powder that we spread with the help of a thin colander cleaned with 70% ethanol. Incubation should be done as sterile as possible, in trays cleaned with 70% ethanol and covered with plastic film as soon as the sample is introduced. Small incisions should be performed for fungal respiration and to prevent condensation in the plastic cover.

2.3 Sensorial analysis

Sensorial analysis of the dishes was performed by the sensory panel of AZTI and Mugaritz according to the flavor profile methodology described by Keane (1992). In essence, 4–5 panelists with broad expertise in sensorial analysis evaluate visual, gustative and flavor characteristics using an open record to annotate their impressions. Sensorial definitions obtained in 2 or 3 different sessions per products were collected, grouped according to their sensorial significance, summarized in consensus descriptors and recorded. Positive and negative descriptors were considered.

3. RESULTS AND DISCUSSION

3.1 Fruits

One of the best substrates for fungal growth is fruits. In this paper, we described the fermentation of apple dices with *R. oryzae*. Ingredients are described in Table 1(prototype A) and the workflow is described in Fig. 1. Briefly, apples (Golden and Granny Smith were used separately in several tests), were diced in cubes of 2 cm, boiled for 60 s in tap water, cooled, dried and fermented for 48 h at 28 °C (measured in the incubator). Time, temperatures and distances between apple dices were the same for both types of apples. After the initial 24 h, fungal filaments could be observed in both apple varieties, and dices were completely covered with *R. oryzae* after 48 h. *Rhizopus* (as all fungus) is an aerobic organism that will colonize the surfaces exposed to air; we, therefore, recommend a minimal distance of 2 cm between apple dices to assure a good coverage. After several trials with both apple varieties, it is observed that the fungus develops faster in more acidic (sensorial) varieties, like the Granny Smith apple. *R. oryzae* would be favored in acidic environments because it counterbalances their own ammonia (and therefore alkalizing) production. It can be also speculated that these varieties favor the fungal growth due to the malic acid concentration differences (Nour et al., 2010), but further experiments are required to confirm this point. Only Granny Smith variety was used for sensorial characterization.

Among the evident visual alterations (seen in Fig. 10), several sensory changes were detected after fermentation comparing raw fruits with fermented. The texture changed from very raw, crunchy, gritty and somewhat fibrous apple dices to dices with a very ephemeral outer layer and succulent interior. Considering flavor, fermentation enhances more sweetness and acidity while eliminating some fruity notes from the reference.

Table 1. List of ingredients for each prototype.

Prototype	Ingredient	Weight/volumen
A	Cooked apple dices of 2 cm	200 g
	Water for cooking	1000 ml
	R. Oryzae starter	1.0 g
B	Cooked rice	200 g
	Water for cooking	1000 ml
	R. Oryzae starter	1.0 g
C	Cooked burghul	200 g
	Water for cooking	1000 ml
	R. Oryzae starter	1.0 g
D	Cooked barley	200 g
	Water for cooking	1000 ml
	R. Oryzae starter	1.0 g
E	Cooked rye	200 g
	Water for cooking	1000 ml
	R. Oryzae starter	1.0 g
F	Cooked nuts	200 g
	Water for cooking	1000 ml
	R. Oryzae starter	1.0 g
G	Cooked peas	200 g
	Water for cooking	1000 ml
	R. Oryzae starter	1.0 g
H	Cooked fava beans	200 g
	Water for cooking	1000 ml
	R. Oryzae starter	1.0 g
I	Cooked beans	200 g
	Water for cooking	1000 ml
	R. Oryzae starter	1.0 g

Externally, products gain a new esthetic appearance with a white and filamentous cover in thereby introducing a new concept for western consumers that is both safe and surprising. The first eye-contact with the product would generate rejection from consumers, and, about 40%

(2 of 5) of the panelists showed reservations about its safety. May be this is because Occidental consumer is not very usual this visual effect in food and it is joined with food in bad conditions to be consumed. This appreciation was observed in all products referred in this paper.

3.2 Grains

Traditionally, sake is obtained through rice fermentation with *Aspergillus oryzae* (Gauntner, 2002). The hereby research presents *R. oryzae* strains as an option to obtain alternative alcoholic products from cooked rice fermentation with particular sensorial characteristics. As occurs with Koji, *R. oryzae* grows best on partially cooked grains and free water should be avoided. This approach is not completely new, because *R. oryzae* is a part of the starters used to prepare other traditional rice wine (Hong Qu glutinous rice wine), (Lv et al., 2012).

R. oryzae was inoculated in different cooked cereals, including rice (*Oryza sativa, L*), burghul (mainly *Triticum durum, L*), barley (*Hordeum vulgare L*) and rye (*Secale cereal, L*). Ingredients are described in Table 1 (rice is prototype B, burghul is prototype C, barley is prototype D, rye is prototype E). The workflow is described in Fig. 2 for rice, Fig. 3 for burghul, Fig. 4 for barley and Fig. 5 for rye. Briefly, grains were boiled separately in a pressure pan until they were completely cooked, mixed with the fungus and then fermented for 48 h at 33 °C. After 48 h, mycelia covered the entire surface exposed to air (Fig. 6).

After the initial 24 h, all tested cereals showed acidity notes with acetic and alcoholic flavors. After another 24 h at refrigeration temperature, the flavor changed to sweet notes. Examples of visual effects with different cereals are shown in Fig. 4. Sensorial panel judged the preparation using rice as a fermentation ingredient as the most acceptable, specially comparing with other fermented cereals. Fermented rice appeared similarly to jelly, like a thick layer of rice-milk pudding, even overcooked rice, without visible mycelium. Initially, the smell was sweet and afterwards appears a sour smell, both pleasant for the sensorial panel. The texture was creamy and melted in the mouth. The flavor was intense, between acidic and sweet, and panelists described it as "rare but not disgusting". The after-taste was less intense and more likeable than in trials with other legumes (results not shown) and grains. Several authors prevent the use of longer fermentation time due the ammonia produced by *Rhyzopus* in these substrates (Steinkraus, 1996).

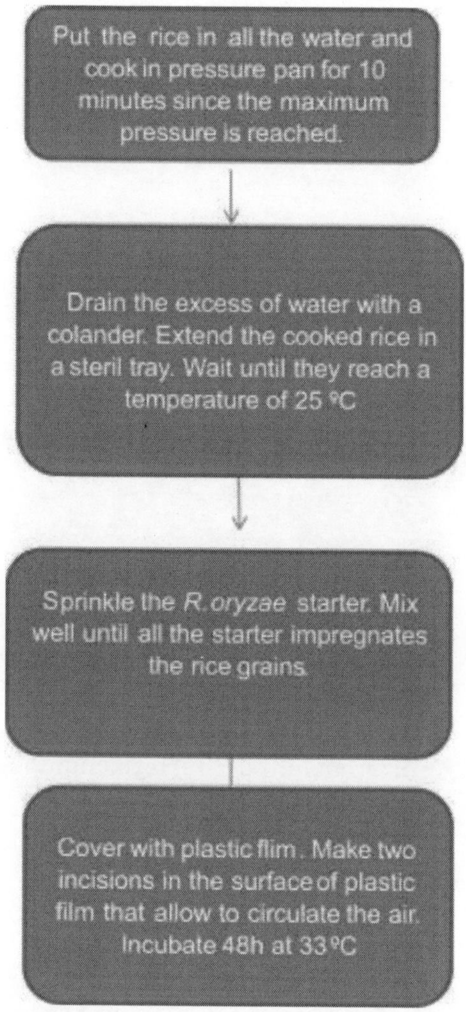

Figure. 2. Workflow diagram for rice fermentation.

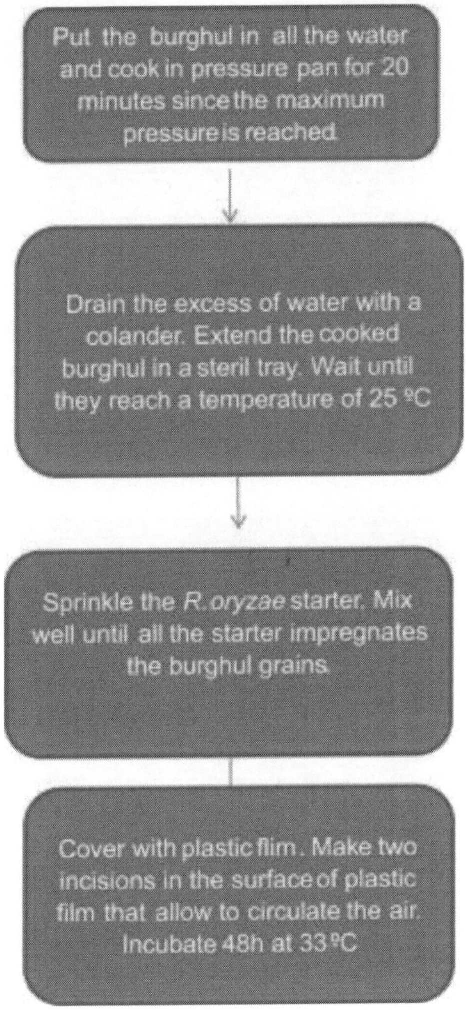

Figure. **3**. Workflow diagram for burghul fermentation.

Other authors have tested different tempeh-like fermentations with grains in order to obtain high-value products. Hachmeister and Fung (1993) fermented hard red winter wheat, triticale, yellow sorghum (milo), and red sorghum with *R. oligosporus*, obtaining only positive sensorial and technological results in the case of red sorghum. In our case, the sensory characteristics of rice after fermentation have completely changed. They were surprising having come from plain cooked rice.

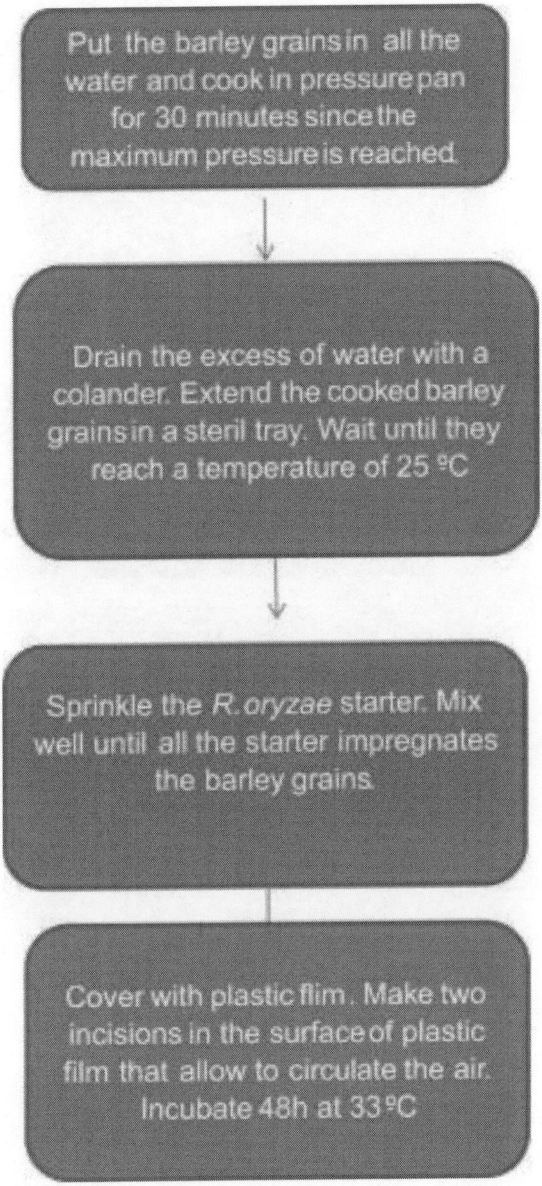

Figure. 4. Workflow diagram for barley fermentation.

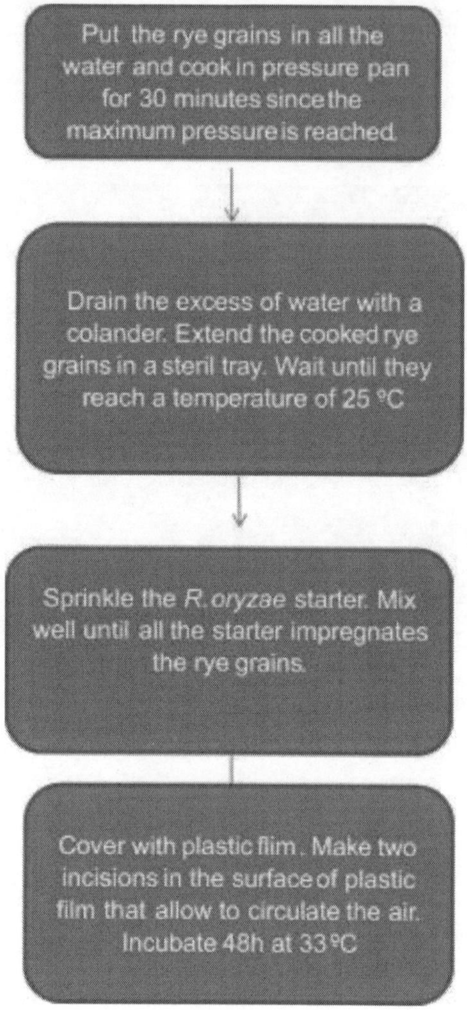

Figure. **5.** Workflow diagram for rye fermentation.

Menu Plate: Cured lobster meat and fermented rice.

Nuts

Following the line of work done with seeds, fruits and vegetables, different tests have also been performed using nuts as fermentations substrates. (Fig. 11)

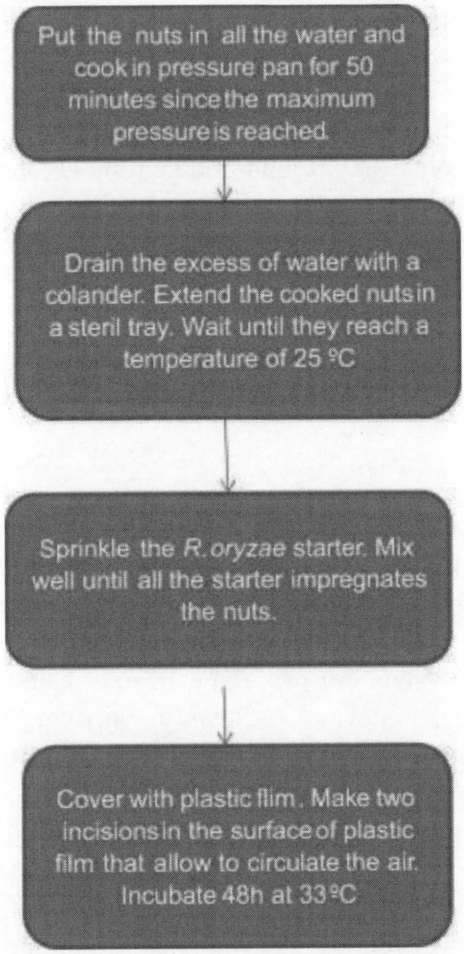

Figure. 6. Workflow diagram for nuts fermentation.

As a general consideration, the fungus required cooked nuts to increase its water activity and make fermentation possible. For example, a preparation with hazelnuts is described in this paper. Ingredients are described in Table 1 (prototype F) and workflow is described in Fig. 5. Briefly, nuts were cooked in a pressure-cooker with water for; they were then drained, tempered, spread on a metal tray and inoculated with fungal starter. The time of incubation lasted 48 h at 33 °C. After this time, the development of mycelium was observed in between individual nuts, sticking them together in a continuous product and forming a structure similar to traditional tempeh (Fig. 12).

The visual aspect of the product was similar to a nougat; with a combination of aromas of hazelnut and light mold. The flavor consisted of cooked hazelnuts, lightly sweetened with notes of honey. The texture was soft, creamy and lightly starchy.

Menu plate: Cooked nougat savory praline and peppercorn.

Legumes

Using the same concept as traditional tempeh (made from cooked soy seeds) *R. oryzae*can be inoculated in other cooked legumes. We tested cooked peas (*Pisum sativum, L*), beans (*Phaseolus vulgaris, L*) and fava beans (*Vacia faba, L*). Ingredients are described in Table 1 (peas are prototype G, fava beans are prototype H and beans are prototype I). Workflow is described in Fig. 7 for peas, in Fig. 8 for beans and in Fig. 9 for fava beans. Briefly, legumes were cooked separately in a pressure cooker with water for 5 min (peas) 10 min (fava beans) and 20 min (beans soaked) since the full pressure was reached (1 bar); they were then drained, tempered, and spread on a metal tray and inoculated with a fungal starter. As it happens in traditional tempeh making, there is an initial bacterial fermentation or acidifying step in legumes when they are soaked before cooked. The time of incubation lasted for 48 h at 33 °C. After this time, mycelium was observed between the legumes, sticking them together in a continuous product and forming a structure similar to traditional tempeh (Fig. 13).

In all cases, visual presence was similar to tempeh, forming a compact and stable block of legumes connected and fused by the fungal hyphae. Molds were visible in the legumes exterior. Flavor was similar to unleavened bread, fungal, with acid and sweet notes. Texture was soft, velvety and juicy. In the three cases, initial taste was acidic, bitter and slightly sweet. As fermentation continues, proteins and lipids hydrolysis promote stronger flavor, and, eventually, ammonia was released. Initial color was bright white, then grey and black because spores are produced by the mold. Similar effect on color was observed in tempeh (Handoyo and Morita, 2006).

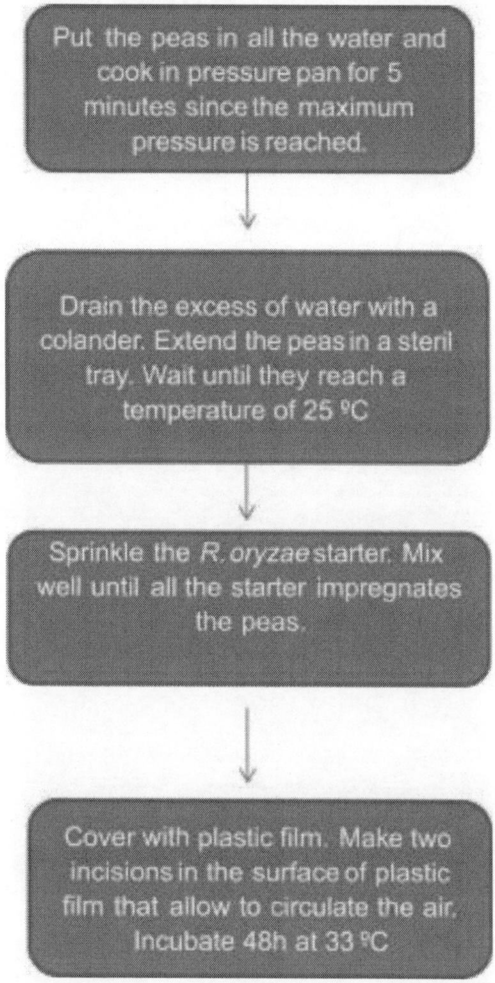

Figure. **7.** Workflow diagram for peas fermentation.

Menu plate: Vegan "alubiada": ***Carpaccio*** **of fermented and toasted Tolosa beans.**

In general, and beyond the initial prejudices, mainly because the unusual external appearance, consumers find a range of complex flavors outside of the conventional and expected. Fermented foods, already present in the menu of the restaurant, has served to be a real test with consumers by checking the degree of acceptability. Further studies would transfer them to the food industry in order to obtain new, healthy and delicious products. This transference would require strains, substrates and procedure adaptation, but would allow a completely new product repertoire.

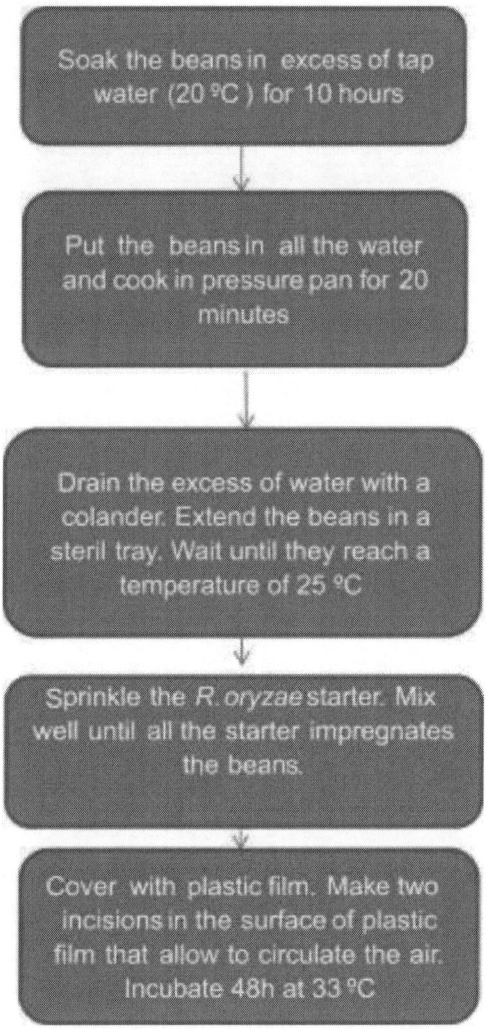

Figure. 8. Workflow diagram for beans fermentation.

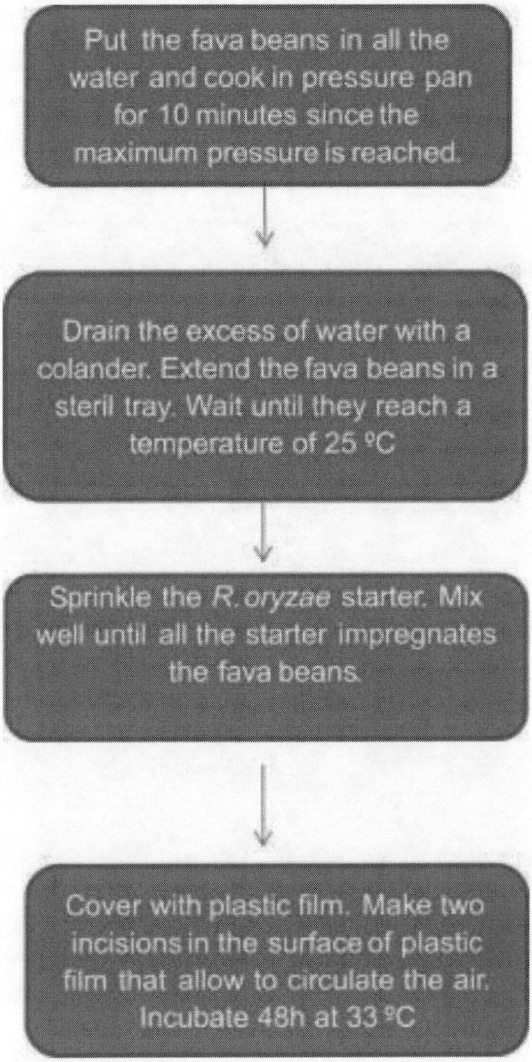

Figure. **9**. Workflow diagram for fava beans fermentation.

4. CONCLUSIONS

Fruits, grains, nuts and legumes mold-fermentation with *R. oryzae* produces profound sensory changes in foods at different levels. New flavors including acidic, sweet and bitter appears in treated products in varying degrees, depending on the fermented food, and modify their sensory profile. The texture is also deeply modified, and, in all cases, visual aspects of fermented products are completely unusual.

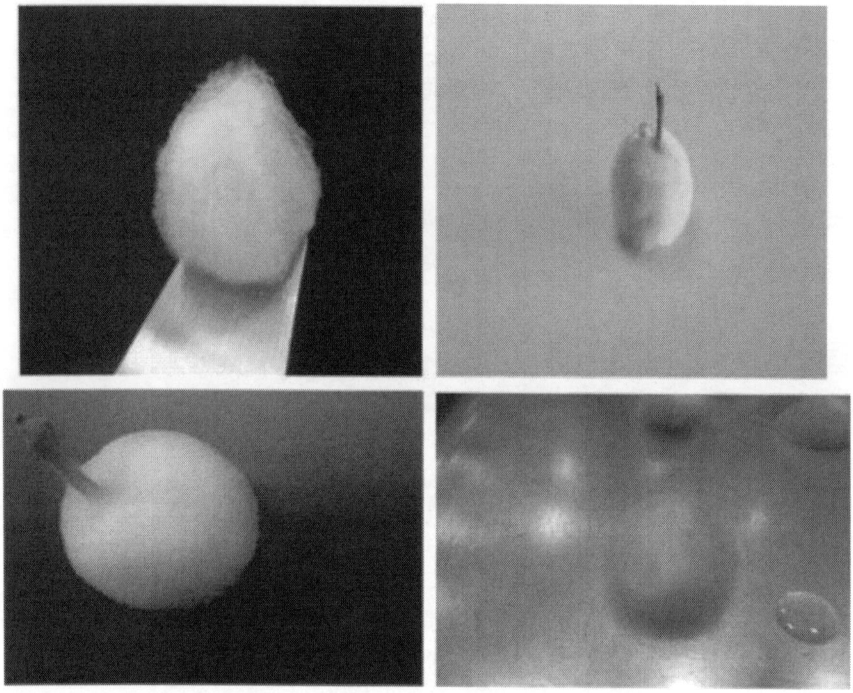

Figure. **10**. Fruits colonized by *R. oryzae* (A) apple dice, (B) whole wild apples, (C) apple puree colonized with *R. oryzae*, and (D).

Figure. 11. Grains colonized by *R. oryzae* (A) rice colonized with *R. oryzae*, (B) burghul colonized with *R. oryzae*, (C) barley colonized with *R. oryzae*, and (D) rye colonized with *R. oryzae*.

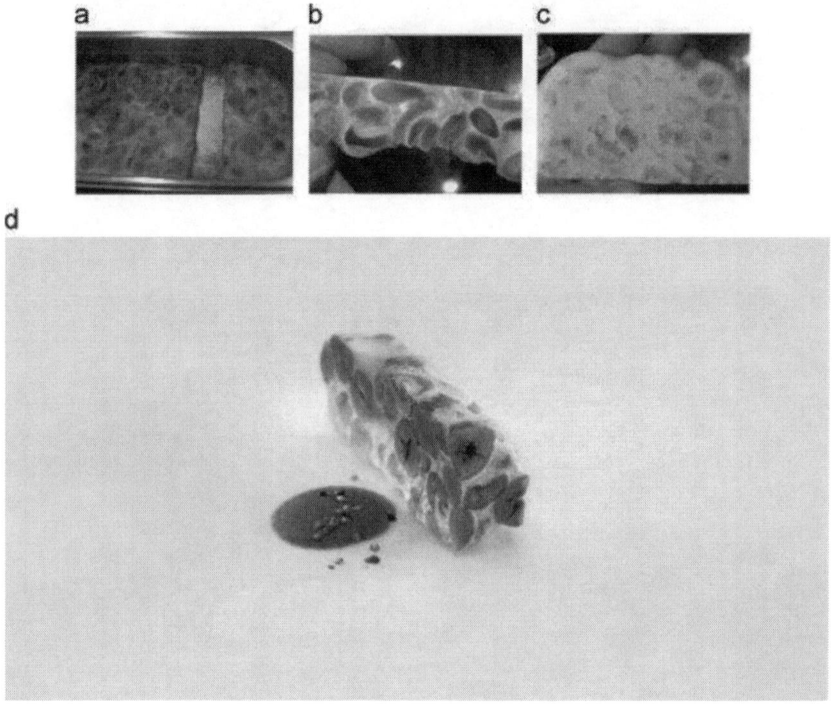

Figure. **12.** (A, B and C) Helznuts colonized with *R. oryzae* and (D) Turrón.

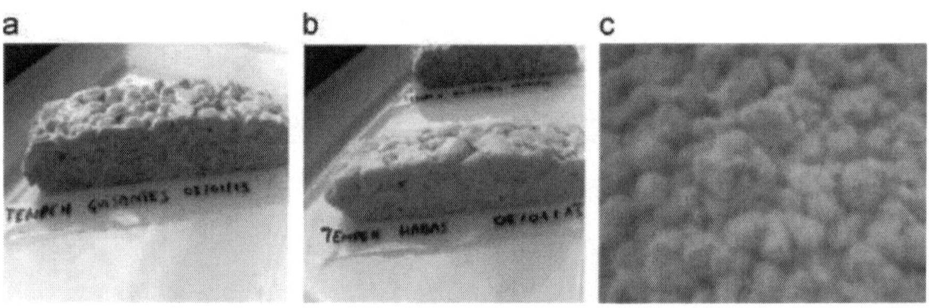

Figure. **13.** Legumes colonized by *R. oryzae* (A) Peas, (B) Broad beans, and (C) beans.

ACKNOWLEDGMENTS

Authors thank the sensorial panel of AZTI-Tecnalia for their excellent technical assistance and patience during this investigation.

REFERENCES

1. Azeke, M.A.,Fretzdorff,B.,Buening-Pfaue, H., Betsche, T., 2007. Compara- tive effect of boiling and solid substrate fermentation using the tempeh fungus (Rhizopus oligosporus) onthe flatulence potentialofAfrican yambean (Sphenostylis stenocarpaL.) seeds. FoodChem.103(4), 1420–1425.

2. Baumann, U.,Bisping, B.,1995. Proteolysis during tempe fermentation. Food Microbiol. 12(0),39–47.

3. Denter, J.,Rehm,H.-J.,Bisping,B.,1998.Changesinthecontentsoffat-soluble vitamins and provitamins during tempe fermentation. Int.J.Food Microbiol. 45(2),129–134.

4. Dinesh Babu, P., Bhakyaraj, R., Vildhyalakshmi, R.,2009. Alowcost nutritious food "tempeh" – A review.WorldJ.DairyFoodSci.4(1), 222–227.

5. Gauntner, J., 2002. Teh Sake Handbook. Tuttle Publishing, Tokyo.

6. Hachmeister, K.A., Fung, D.Y.C.,1993. Tempeh: amold-modified indigenous fermented food made from soybeans and/ orcerealgrains.Crit.Rev. Microbiol. 19(3),137–188.

7. Handoyo, T.,Morita,N.,2006.Structuralandfunctionalpropertiesof fermented soybean(tempeh)by using Rhizopus oligosporus. Int. J.Food Prop. 9,347–355.

8. Keane, P.,1992.The flavor profile. In:Hootman,R.C.(Ed.),Manualon Descriptive Analysis Testing for Sensory Evaluation 1sted.ASTM, Philadelphia, USA,pp.5–14.

9. Lv, X.-C.,Huang,Z.-Q.,Zhang,W.,Rao,P.-F.,Ni,L.,2012.Identification and characterizationof filamentous fungiisolatedfromfermentation starters for Hong Quglutinousricewinebrewing.J.Gen.Appl.Microbiol. 58 (1),33–42.

10. Nour, V.,Trandafir, I.,Ionica, M.E., 2010. Compositional characteristics of fruits of several apple (Malus domestica Borkh.) cultivars. Not. Bot. Horti Agrobot. Cluj-Napoca38(3),228–233.

11. Shurtleff, W., Aoyagi, A., 2001. The Book of Tempeh. Ten Speed Press.

12. Steinkraus, K.H., 1996. Handbool of Indigenous Fermented Foods, 2nd ed. Marcel Dekker Inc., New York.

13. Varga, J.,Peteri,Z.,Tabori,K.,Teren,J.,Vagvolgyi,C.,2005.Degradationof ochratoxin AandothermycotoxinsbyRhizopusisolates.IntJ.Food Microbiol. 99(3),321–328.

14. Wiesel, I.,Rehm,H.J., Bisping,B., 1997. Improvement of tempe fermentations by application of mixed cultures consisting of Rhizopus sp. andbacterial strains. Appl. Microbiol. Biotechnol. 47(3),218–225.

13

COOKING AND NUTRITIONAL SCIENCE: GASTRONOMY GOES FURTHER

Virginia Navarro[a], Gema Serrano[a, b], Dani Lasa[b], Andoni Luis Aduriz[b], Josune Ayo[a]

[a] AZTI—Tecnalia Food Research Institute, Biscay Technology Park, Astondo Bidea, Building 609, 48160 Derio, Bizkaia, Spain

[b] Mugaritz Restaurant, Otzazulueta Baserria, Aldura Aldea 20 zk, Errenteria, 20100 Gipuzkoa, Spain

ABSTRACT

Science-based cooking is closely associated with the design of stimulating and novel dishes that make guests feel an explosion of sensations. Chefs are expected to use high quality foods and thorough preparation techniques. But food science is not only texture and technology, it is also nutrition and health. From a nutritional point of view, science-based cooking may contribute to providing certain nutrients and other food components, which could confer healthy aspects to the dishes and menus. Chefs may then also consider nutritional aspects when designing dishes and menus. The purpose of the present study was to evaluate the nutritional profile of the innovative dishes and menus offered in Mugaritz Restaurant in Spain. European food legislation and recommendations have been applied in order to evaluate two menus and to globally understand the impact they have on diet and health.

KEYWORDS

Gastronomy; Nutritional profile; Health properties

1. INTRODUCTION

The evolution and history of gastronomy has been clearly influenced by the social and economic background in each period of time. In 1825, gastronomy was defined as 'the reasoned knowledge of all that relates to man feeding himself'. Its aim was to 'attend to the preservation of man by means of the best possible food' (Vega and Ubbink, 2008). In the past, the aim of a diet was to provide nutritional requirements in order to avoid nutrient deficiencies. This aim has evolved into the desire to keep or improve health status through the diet. Likewise, the evolution of gastronomy into haute cuisine has led to the utilization of new ingredients and technologies that could interact with nutrients and alter the contribution of the dishes to the overall diet. The main goal of haute cuisine cooks is to innovate and design delicious dishes with new textures and flavors, which promote new sensations in customers. The logical next step of this goal for haute cuisine could be to understand the impact of these new processes and ingredient mixtures on the nutritional composition of the dishes and menus, and furthermore, the impact of the menus on the overall diet of the individual. Cooks could implement this knowledge in the design of new dishes in order to equilibrate the menus in terms of nutritional and health properties. As gastronomy and haute cuisine are intensely related to social interactions, it is extremely interesting to analyze their implications on one of the major current concerns of society: health. In this sense, European Food Safety Authority (EFSA) is concerned about new product launches with nutritional and health claims that need to be substantiated. New legislation is consequently being put together in order to control and define the claims made by the manufacturers of these products. Bearing these indications from EFSA in mind, the results presented here are discussed in terms of nutritional and health properties of two different menus from Mugaritz Restaurant.

2. MATERIALS AND METHODS

2.1 Menu design

The working team of the experimental kitchen from Mugaritz Restaurant was responsible for defining and designing the dishes. The selection of the final recipes and menus was made attending to guest demands and sensory preferences. Recipes are described inSupplementary material. Final presentation of a dish is shown in the figures, for example,*A pasta of amaranth, baby langoustine tails,* is shown in Fig. 1.

Figure. 1. Presentation of the dish of *A pasta of amaranth, baby langoustine tails. Tender garden leaves.* Photograph from Jose Luis López de Zubiria.

2.2 Nutritional composition analysis of dishes and menus

A total of 21 dishes conforming to two different menus (Sustraiak and Naturan) were analyzed. All dishes were prepared in duplicate and weighed. Each dish was then completely homogenized and a representative sample was used for nutritional composition analysis. Macro and micronutrients as well as certain bioactive compounds were determined. Samples were analyzed according to the Association of Official Analytical Chemists methods (AOAC, 2002): moisture (AOAC 950.46), ash (AOAC 923.03) and protein content (AOAC 954.01). The measurement of reducing sugars was expressed as invert sugar or dextrose equivalent by the Luff–Schoorl method (Lees, 1968).

Total lipids were evaluated by extraction with chloroform/methanol/water following the Bligh & Dyer method with minor modifications (Bligh and Dyer, 1959). The quantitative cholesterol content of the samples was determined according to the Annex V of the European Official Analysis Methods (European Commission, 1991a). Quantification of the silanised sterol fraction was carried out by capillary gas chromatography, on a Hewlett Packard 6890 chromatograph with flame ionization detector (FID) using a SE-54 capillary column (30 m×0.25 mm i.d. and 0.25 μm thickness, Supelco). Lipid extractions from each sample were esterified as described in Annex X of the European Official Analysis Methods (European Commission, 1991b). Fatty acid methyl esters (FAMEs) were analyzed using a gas chromatograph Hewlett Packard 5860 equipped with a flame ionization detector 6890 (FID) and a DB-23 fused silica capillary column (60 m×0.25 mm i.d. and 0.25 μm thickness; Agilent Technologies, USA). Fatty acids were identified by comparison of retention times to standards and relative quantities were expressed as weight percent of total fatty acids.

Total dietary fiber (TDF) was analyzed following the official enzymatic–gravimetric method (AOAC 991.43), total sodium (Na) and potassium (K) by a flame photometric method (AOAC 969.23), calcium (Ca) by atomic absorption spectrophotometry (AOAC 975.03) in a fast sequential atomic absorption spectrophotometer (model AA240 FS, Varian Inc., The Netherlands), following the official methods (AOAC, 2002). The analytical curve was plotted for each element. All the nutritional composition analyses were done in duplicate and the results shown in the tables are the means.

2.3 Nutrition and health assessment tools

For the nutrition and health assessment of Mugaritz food designs, the current legislation on nutrition and health claims of foodstuffs was utilized. As a consequence, the nutrients of compulsory declaration for nutritional claims were analyzed. Total energy was also calculated as indicated by law: energy (kcal)=4×(g protein+g carbohydrate)+9×(g lipid)+2×(g of TDF) (European Commission, 2011).

In order to assess the real contribution of each dish and menu to the diet, results were expressed considering the servings given to the customers in the restaurant. This information allowed the comparison between nutritional values of all the dishes (per serving) and evaluation of their contribution to the menu.

In order to convey the relative significance of the contribution of each dish and menu to a daily diet, the results were also expressed as a percentage of a Daily Reference Value (DRV) or Reference Intake (RI). DRVs or RIs for energy and selected nutrients were those proposed by the Panel on Dietetic Products, Nutrition and Allergies of the European Food Safety Authority (EFSA) for a woman (European Food Safety Authority (EFSA), 2009a and European Food Safety Authority (EFSA), 2010a).

In order to highlight their nutritional properties or potential effects on health, it was also necessary to evaluate the overall nutritional status of the dish and menu designs. To that end, dishes were categorized on the basis of their composition according to a legal proposal of the European Commission (2009). For each food category, thresholds of sodium, saturated fat and sugars have been suggested. Dishes complying with these nutritional criteria would be justified in making nutrition and health claims according to food regulations (European Commission, 2006, European Commission, 2010 and European Commission, 2011). Thus only those dishes, which did not exceed EFSA thresholds, were evaluated for nutritional and health claims.

Additionally, the presence of some nutrients (sodium and cholesterol) and also certain bioactive components (long chain omega-3, linolenic acid and potassium) involved in the etiology (formation or prevention) of cardiovascular diseases were evaluated. Finally, based on scientific opinions of EFSA regarding the health aspects of food components, menus were assessed for their potential health benefit for the general population.

3. RESULTS AND DISCUSSION

3.1 Nutritional profile of dishes and menus

3.1.1 Nutritional composition per serving

The basic nutritional composition of the dishes, expressed by portion, was assessed focusing on nutrients of compulsory declaration according to the Regulation 1169/2011 (European Commission, 2011). Table 1 and Table 2 summarize the results for 21 dishes divided into two different menus (Sustraiak and Naturan).

In general, the nutritional quality of the dishes (in terms of energy, fat, sugar and sodium) was remarkable and appropriate for a healthy diet, with some exceptions. As shown inTable 1 and Table 2, those dishes that provided the highest energy, total fat and saturated fat were *Caramelized sweet French toast, Iberian suckling pig, Euskal Herria cheeses*(from the Sustraiak menu) and *Foie-Gras with mustard* and *Braised Iberian pork tails and pan fried langoustines* (from the Naturan menu). As fat provides more calories per gram than other nutrients, these dishes, which are high in fat, contribute more importantly to total calories. Although the Sustraiak menu had more dishes with a low-calorie content than the Naturan, three high-calorie dishes from Sustraiak brought total calorie content and saturated fat content for both menus to the same level.

Table 1. Intake of the critical nutrients per serving of each dish from the Sustraiak menu.

Dish[a]	Calories (kcal)	Protein (g)	Carbohydrates (g)	Total sugars (g)	Fat (g)	SFA[b] (g)	TDF[c] (g)	Sodium (mg)
Ravioli filling of crab and fresh chestnuts	28.51	4.36	1.94	0.39	0.22	Tr	0.68	444.09
Mini mozzarellas: buttery Idiazabal cheese gnocci in salted Iberian pork bouillon	25.45	0.87	3.23	1.12	0.86	0.24	0.68	111.63
A pasta of amaranth, baby langoustine tails	94.78	12.22	4.09	0.12	2.35	0.66	4.18	271.55
Honeyed fish stew	88.90	18.94	0.32	0.85	1.23	0.35	0.41	638.47
Iberian suckling pig	284.19	19.39	8.84	0.24	18.94	5.45	0.41	432.79
Selection of Euskal Herria cheeses	214.28	12.28	10.61	10.24	12.06	7.34	7.09	426.97
Carrots, hazelnut cream and gomasio	37.57	0.60	6.28	6.46	0.85	0.20	1.23	42.54
Caramelized sweet French toast	258.63	7.12	26.75	32.50	13.33	7.90	1.62	171.47
Edible stones (boiled potatoes in gray clay)	56.82	1.04	3.49	1.30	4.15	0.84	0.69	37.67
White fried shrimp	37.66	2.73	0.15	0.02	2.63	0.29	1.21	144.30
Organic wheat bread	173.97	5.19	36.42	5.36	0.26	0.05	2.58	426.20
Menu	**1300.75**	**84.74**	**102.11**	**58.59**	**56.87**	**23.32**	**20.76**	**3147.70**

[a]Complete information of dishes is detailed in the Supplementary material.
[b]Saturated fatty acids.
[c]Total dietary fiber.

Table 2. Intake of the critical nutrients per serving of each dish from the Naturan menu.

Dish[a]	Calories (kcal)	Protein (g)	Carbohydrates (g)	Total sugars (g)	Fat (g)	SFA[b] (g)	TDF[c] (g)	Sodium (mg)
Vegetable carpaccio	147.45	4.86	8.97	5.62	9.86	2.55	1.66	107.91
Silky kokotxa of bacalao	41.80	5.64	0.36	0.24	1.83	0.38	0.69	301.72
Heart of baby leeks	53.96	5.08	2.02	1.68	2.37	0.55	2.10	795.43
Foie-Gras with mustard	236.42	6.79	4.47	1.17	20.26	7.22	4.53	291.22
Loin of sole	167.77	17.33	3.99	0.73	8.98	4.18	0.83	457.80
Sautéed red mullet fillets	122.36	13.97	2.49	1.25	5.65	1.85	2.84	611.34
A piece of veal	121.34	20.16	5.65	1.55	1.72	0.74	1.35	841.28
Braised Iberian pork tails and pan fried langoustines	231.80	22.61	4.17	2.44	13.63	4.46	1.02	426.93
Warm squash	85.83	14.75	2.59	3.99	1.58	1.04	1.10	77.98
Interpretation of vanity: moist chocolate cake, cold almond cream, bubbles, smoke and cocoa	93.37	2.08	5.69	3.91	6.57	4.58	1.56	17.81
Edible stones (boiled potatoes in gray clay)	56.82	1.04	3.49	1.30	4.15	0.84	0.69	37.67
White fried shrimp	37.66	2.73	0.15	0.02	2.63	0.29	1.21	144.30
Organic wheat bread	173.97	5.19	36.42	5.36	0.26	0.05	2.58	426.20
Menu	**1381.30**	**111.72**	**71.13**	**23.38**	**67.80**	**25.79**	**19.83**	**4127.95**

[a]Complete information of dishes is detailed in the Supplementary material.
[b]Saturated fatty acids.
[c]Total dietary fiber.

The highest protein content was seen in Honeyed fish stew and Iberian suckling pig from the Sustraiak and A piece of veal, Braised Iberian pork tails and pan fried langoustines and Loin of sole from the Naturan. All of these were composed mainly of meat or fish, thus naturally rich in this nutrient.

Sugar content in the Sustraiak menu was higher than in the Naturan, basically caused by the high sugar content of Caramelized sweet French toast (32.50 g), a dessert rich in sugar, cream and butter. Moreover, Selection of Euskal Herria cheeses was the dish with the second highest sugar content (10.24 g), and it was in the Sustraiak menu as well. This notably increased the difference in total sugar content between the two menus.

The Naturan menu was richer in sodium than the Sustraiak, mainly because of two dishes: Heart of baby leeks and A piece of veal (795.43 and 841.28 mg, respectively). Thus, even though the rest of dishes were similar in terms of sodium content, the Naturan menu had two more dishes, which, when summed, accounted for the total sodium content difference between the menus.

Regarding the total dietary fiber, it was remarkable that there were high differences among dishes but the fiber content of the menus was quite similar. As an example, three dishes (Iberian suckling pig, Selection of Euskal Herria cheeses and Caramelized sweet French toast) from the Sustraiak menu, contributed almost for the 90% of the TDF.

Considering the contribution of the menus to the total diet, it is important to highlight that menus were not fully equilibrated because of their lack of carbohydrates. In this study the breakfast was not evaluated, so this meal would compensate their lack on the diet because it is usually the major supplier of carbohydrates. On the other hand, fat, protein and sodium content were above the recommendations. The appropriate layout of some critical dishes could easily help to nutritionally equilibrate the menus.

3.1.2 Contribution of the menus to Dietary Reference Values and Reference Intakes

Table 3 and Table 4 reflect the contribution of each menu to the DRVs, defined in the Materials and methods section. This information was very useful to understand the real impact of each dish and menu on the global diet.

The Naturan menu had a stronger impact on total calories, total fat and total saturated fat recommendations than the Sustraiak menu, as was expected. Neither menu exceeded total daily recommended calories, but the Naturan menu reached the 80th percentile. It is necessary here to highlight that both fat and saturated fat in the Naturan menu exceeded recommendations (113.56% and 143.60%, respectively). It is noticeable that for the dish Interpretation of vanity, its contribution to recommended fat intake is not as high as its contribution to saturated fat intake. The use of butter and other lipids rich in saturated fat like cocoa could explain this fact. Saturated fat was also excessive in the Sustraiak menu, and in this case, one dish contributed more than the others: Caramelized sweet French toast. It is clearly something to be aware of, because the intake of saturated fat is one of the most important risk factors for cardiovascular diseases and obesity.

Sugars did not reach the DRV but were clearly higher in the Sustraiak menu because of the dish Caramelized sweet French toast. Salt has one of the most restrictive recommendations for cooking. It is a key ingredient because of its implication in cardiovascular and renal diseases but also because of its sensorial properties. It is easy to reach the 6 g Reference Intake (RI) of salt published by EFSA. Nevertheless, we tried to identify dishes with excessive sodium content in order to assess the addition of salt. There were three major contributors to total sodium content of menus: A piece of veal, Sautéed red mullet fillets and Honeyed fish stew, all of them more than 20% above the RI. Two of these were located in the Naturan menu, which exceeded the Reference Intake for sodium more than the Sustraiak menu (192% and 133%). The total sodium content of both menus seems a critical nutrient to consider. Although it is easy to control the salt content of the dishes, it should be considered that salt is not the unique sodium source.

Table 3. Percentage of the Dietary Reference Values (DRVs) and References Intakes (RIs) covered in Sustraiak menu.

Dish[a]	Calories	Fat	SFA[b]	Carbohydrates	Sugars	Sodium	TDF[c]
Ravioli filling of crab and fresh chestnuts	1.43	0.31	Tr	0.84	0.44	18.82	2.70
Mini mozzarellas: buttery Idiazabal cheese gnocci in salted Iberian pork bouillon	1.27	1.22	1.18	1.40	1.25	4.73	2.71
A pasta of amaranth., baby langoustine tails	4.74	3.36	3.31	1.78	0.14	11.51	16.71
Honeyed fish stew	4.44	1.75	1.74	0.14	0.94	27.05	1.62
Iberian suckling pig	14.21	27.06	27.26	3.84	0.26	18.34	1.64
Selection of Euskal Herria cheeses	10.71	17.23	36.70	4.61	11.37	18.09	28.35
Carrots, hazelnut cream and gomasio	1.88	1.21	1.02	2.73	7.18	1.80	4.91
Caramelized sweet French toast	12.93	19.04	39.49	11.63	36.11	7.27	6.47
Edible stones (boiled potatoes in gray clay)	2.84	5.93	4.21	1.52	1.44	1.60	2.76
White fried shrimp	1.88	3.76	1.44	0.07	0.02	6.11	4.85
Organic wheat bread	8.70	0.37	0.27	15.84	5.96	18.06	10.33
Menu	**65.04**	**81.24**	**116.61**	**44.40**	**65.10**	**133.38**	**83.05**

[a]Complete information of dishes is detailed in the Supplementary material.
[b]Total dietary fiber.
[c]Saturated fatty acids.

Table 4. Percentage of the Dietary Reference Values (DRVs) and References Intakes (RIs) covered in Naturan menu.

Dish[a]	Calories	Fat	SFA[b]	Carbohydrates	Sugars	Sodium	TDF[c]
Vegetable carpaccio	7.37	14.09	12.74	3.90	6.24	4.57	6.66
Silky kokotxa of bacalao	2.09	2.61	1.91	0.15	0.27	12.78	2.76
Heart of baby leeks	2.70	3.39	2.76	0.88	1.87	33.70	8.40
Foie-Gras with mustard	11.82	28.94	36.08	1.94	1.30	12.34	18.13
Loin of sole	8.39	12.83	20.92	1.74	0.81	19.40	3.33
Sautéed red mullet fillets	6.12	8.07	9.23	1.08	1.38	25.90	11.38
A piece of veal	6.07	2.45	3.68	2.46	1.72	35.65	5.39
Braised Iberian pork tails and pan fried langoustines	11.59	19.47	22.29	1.81	2.71	18.09	4.08
Warm squash	4.29	2.26	5.21	1.13	4.43	3.30	4.42
Interpretation of vanity: moist chocolate cake., cold almond cream., bubbles., smoke and cocoa	4.67	9.39	22.88	2.47	4.35	0.75	6.25
Edible stones (boiled potatoes in gray clay)	2.84	5.93	4.21	1.52	1.44	1.60	2.76
White fried shrimp	1.88	3.76	1.44	0.07	0.02	6.11	4.85
Organic wheat bread	8.70	0.37	0.27	15.84	5.96	18.06	10.33
Menu	**78.53**	**113.56**	**143.60**	**34.98**	**32.49**	**192.27**	**88.73**

[a]Complete information of dishes is detailed in the Supplementary material.
[b]Saturated fatty acids.
[c]Total dietary fiber.

Total dietary fiber was near 90% in both menus, which is extremely positive. The general population does not consume the daily recommended amount of fiber (25 g), and the use of cellulose and other fibers in the design of restaurant dishes would help to reach this recommendation.

3.1.3 Classification of dishes according to the legal proposal from the European Commission (EC)

In order to assess the nutritional status of the dishes, the legal proposal from the EC was applied. As mentioned in the Materials and methods section, a legal preliminary draft proposal was published in 2009 (European Commission, 2009). This draft defined the highest acceptable sodium, saturated fat and total sugars content for certain categories of food taking into account their contribution to the diet but, it is important to point out that this preliminary proposal is not definitive. Also, it was not easy to decide which of the food categories in the proposal were adequate for such elaborate dishes as those considered in this study and, in some cases the classification may not be the best considering the characteristics of the dishes (Table 5). As an example, Caramelized sweet French toast and Interpretation of vanity were classified in the cereals group although this group has a fat content limit that most desserts could not meet (and there is not a dessert category).

After the classification of the Mugaritz dishes into groups, thresholds in sodium, saturates and sugars were checked. Dishes not satisfying these limits were not considered for further health and nutritional claims. Results are shown in Table 5. Values over the thresholds proposed by the EC are given in bold font.

Table 5. Classification of the dishes by categories of food. Evaluation based on the critical nutrients.

Dish[a]	Classification	Recommendation for the critical nutrients			Nutrient profiles		
		Total sugars (g/100 g)	Sodium (mg/100 g)	SFA[b] (g/100 g)	Total sugars (g/100 g)	Sodium (mg/100 g)	SFA[b] (g/100 g)
Caramelized sweet French toast	Cereals and cereal products. Breakfast cereals	25	500	5	**25.00**	131.90	**6.08**
Interpretation of vanity: moist chocolate cake. Cold almond cream. Bubbles. smoke and cocoa					8.89	40.47	**10.40**
Selection of Euskal Herria cheeses	Dairy based products. Cheeses	15	600	10	7.76	323.46	5.56
A pasta of amaranth., baby langoustine tails	Fish. Fishery products. Crustaceans and molluscs	–	700	10	0.20	445.17	1.09
Honeyed fish stew					0.95	**717.38**	0.39
Silky kokotxa of bacalao					0.61	**754.30**	0.96
Loin of sole					0.62	387.97	3.55
Sautéed red mullet fillets					1.25	611.34	1.85
White fried shrimp					0.10	**962.03**	1.92
Iberian suckling pig	Meat or meat based products	–	700	5	0.20	363.69	4.58
Foie-Gras with mustard					1.41	350.86	**8.70**
A piece of veal					1.55	**841.28**	0.74
Braised Iberian pork tails and pan fried langoustines					2.32	406.60	4.25
Ravioli filling of crab and fresh chestnuts	Ready meals. Soups and sandwiches	10	400	5	0.20	225.43	Tr
Mini mozzarellas: buttery Idiazabal cheese gnocci in salted Iberian pork boullion					2.04	202.97	0.43
Vegetable carpaccio	Fruits. vegetables and their products. Except oils	15	400	5	5.85	112.41	2.66
Carrots. Hazelnut cream and gomasio					**23.92**	157.56	0.76
Heart of baby leeks					1.83	**864.59**	0.60
Warm squash					5.95	116.39	1.56
Edible stones (boiled potatoes in gray clay)					3.94	114.16	2.55
Organic wheat bread	Cereals and cereal products. Breads	15	700	5	8.00	636.12	0.08

[a]Complete information of dishes is detailed in the Supplementary material.
[b]Saturated fatty acids.

12 Dishes of 21 satisfied all the limits of its group. Briefly, in 5 dishes, sodium content should be slightly controlled, especially in the fish and fishery products group. Saturated fat content was unacceptable only in 3 dishes, two of them being desserts and the last one foie-gras

(naturally rich in saturated fat). In the case of total sugars, they were not appropriate for Caramelized sweet French toast and Carrots, hazelnut cream and gomasio. The limit was not excessively crossed in the case of Caramelized sweet French toast. Carrots, hazelnut cream and gomasio was classified as fruits and vegetables, whose threshold in sugar content is very low for a dish that is served as a dessert. Nevertheless, the high sugar could be mentioned as a warning point.

3.2 Nutritional and health claims

Foods with a particularly beneficial nutritional property (regarding nutrient profile and provided energy) and foods that have demonstrated a relationship with any health aspect can carry a claim of the mentioned positive properties. In the context of the EU Regulation, food bearing any such claim should not have an undesirable nutrient composition, in particular regarding those critical nutrients, salt, fat and sugar, for which excessive intakes are not recommended (European Commission, 2009). Thus, after the overall assessment of nutritional status of the dishes, those complying with key nutrient thresholds were checked for nutritional and health claims. Nutritional claim specifications according to Regulation (EC) 1924/2006 and 116/2010 were applied to dishes, which met these criteria. Allowed nutritional claims are shown in Table 6.

All 12 dishes complying with the specific nutrient profile could claim some nutritional properties. Two of them had up to nine possible nutritional claims (Ravioli filling of crab and fresh chestnuts and A pasta of amaranth and baby langoustine tails). The most frequently allowed claims were those related to protein content. In most of the dishes, sugar and fiber content were low or high enough, respectively, to bear claims. Dishes that contributed to the overall high sugar content of menus were discarded previously, so the vast majority of dishes here could bear the Low sugars claim. The less frequent claims were Low energy, Saturated fat-free and Source of calcium, with only one dish allowed for each of these claims. Potassium levels were analyzed but were not high enough to bear the high potassium claim in any of the dishes.

Table 6. Nutritional claims for filtered dishes.

Dish[a]	Low energy	Low fat	Fat-free	Low saturated fat[b]	Saturated fat-free[b]	Low sugars	Sugars-free	With no added sugars	Low sodium/ salt	High fiber	Source of fiber	High protein	Source of protein	Source of calcium	Source of omega-3 (EPA + DHA)	High monounsaturated fat	High unsaturated fat
Ravioli filling of crab and fresh chestnuts	•	•	•			•	•	•			•	•	•				
Mini mozzarellas: buttery Idiazabal cheese gnocci in salted Iberian pork bouillon		•		•		•		•			•		•				•
A pasta of amaranth., baby langoustine tails				•		•	•	•		•	•	•	•		•		
Iberian suckling pig						•	•	•				•	•			•	
Selection of Euskal Herria cheeses										•	•	•	•	•			
Vegetable carpaccio								•	•				•			•	•
Loin of sole						•						•	•		•		
Sautéed red mullet fillets						•		•			•	•	•		•	•	
Braised Iberian pork tails and pan fried langoustines						•		•				•	•			•	
Warm squash		•							•			•	•				
Edible stones (boiled potatoes in gray clay)						•			•							•	•
Organic wheat bread		•	•	•	•			•			•						

[a]Complete information of dishes is detailed in the Supplementary material.
[b]Trans-fatty acids were not determined.

Although the majority of the dishes showed strong nutritional properties, enough for making validated claims, the total composition of the menus must also be considered. In order to make any claim, the entire menu should be nutritionally equilibrated, as previously mentioned.

Regarding health claims, based on recently published scientific opinions by EFSA and data available from menus, we tried to determine if any of the dishes could bear health claims.

All dishes containing at least enough protein to bear the nutritional claim "source of protein" could also bear the health claim "protein contributes to the maintenance of bone" and "protein contributes to the growth or maintenance of muscle mass" (EFSA, 2010b). As shown in Table 6, this occurred for almost all dishes considered (10 of a total of 12).

Similarly, all dishes containing at least enough calcium to bear the nutritional claim "source of calcium" could also bear the following health claims: "calcium is needed for the maintenance of normal bones and teeth", "calcium contributes to normal muscle function and neurotransmission", "calcium contributes to normal blood clotting", "calcium contributes to normal energy metabolism" and "calcium contributes to the normal function of digestive enzymes" (EFSA, 2010c). This was the case for Selection of Euskal Herria cheeses.

In order to bear the claim alpha-linoleic acid (LA) or alpha-linolenic acid (ALA) helps to maintain normal blood cholesterol concentrations; dishes should have at least 1.5 and 0.3 g of each fatty acid, respectively (European Food Safety Authority (EFSA), 2009b and European Food Safety Authority (EFSA), 2009c). White fried shrimp was rich enough in alpha-linoleic acid to bear a claim (data not shown), but it was not considered for nutritional and health claims because of its high sodium content. Two other dishes could bear a claim based on their fatty acid profiles: Iberian suckling pig, because of its alpha-linoleic acid content (2.31 g per serving) and Loin of sole for its alpha-linolenic acid content (0.4 g per serving).

It is also interesting to evaluate the total fatty acid intake for the menus as a whole because health claims could be justified when intake levels are achieved overall. Logically, taking this into account, both menus could bear the previously mentioned claims. Moreover, the Naturan menu could bear the claim related to eicosapentaenoic acid (EPA) and EPA+docosahexaenoic acid (DHA) levels (440.90 and 638.45 mg, respectively). In this case, the health claims are as follows: "its DHA levels contribute to the maintenance of normal brain function and to the maintenance of normal vision" (EFSA, 2010d) and "its EPA+DHA levels contribute to the normal function of the heart" (EFSA, 2010e). Nevertheless, it is important to point out that claims for the menus as a whole require taking into account dishes that were discarded because they did not meet the EC draft requirements for threshold values of key nutrients.

There is not any health or nutritional claim regarding low levels of cholesterol. Nevertheless it is interesting to analyze because this nutrient is essential for life but, at the same time, its serum levels are an important biomarker for cardiovascular diseases. There was not an excessive cholesterol content in any of the menus (138.66 mg for Sustraiak and 141.58 mg for Naturan), even considering the recommendation for persons whose LDL levels are above the goal (<200 mg of cholesterol intake per day) (NECP, 2002).

Some isolated dishes could bear the nutritional claim low sodium/salt (Vegetable carpaccio, Warm squash and Edible stones). These dishes could then bear the health claim "High sodium intake increases blood pressure; consumption of foods low or very low in sodium helps to maintain normal blood pressure" (EFSA, 2011). As has been previously described, total sodium content of both menus should be taken into account.

4. CONCLUSION

In general, when focusing on the maximum recommended amount of critical nutrients such as saturated fat, sugar and sodium, more than half of the dishes from both menus were nutritionally well-balanced. Those dishes that exceeded the suggested criteria for the critical nutrients were studied to identify the responsible ingredients as an opportunity for the improvement of their nutritional quality and for the improvement of dishes and menus in general. In addition, it was found that nutritional claims could be made for all the dishes, which did not exceed the suggested criteria. Furthermore, some dishes had significant amounts of certain bioactive compounds, which may help in the prevention of cardiovascular diseases. Others could bear some health claims regarding their protein and calcium content.

Summing up, fine dining and cooking, in general, have become an important part of leisure time. Haute cuisine has gotten closer to society and vice versa and consequently, both of them should be aware of the others' motivations and concerns. This approach or symbiosis has changed the way of cooking in most restaurants, forcing an adaptation to the desires and necessities suggested by the most heterogeneous gourmet community in history.

The sensitivity that Mugaritz has developed in matters of health and nutrition is as important as any given feedback received directly from their customers. Without giving up the creative and artisan component that characterizes restaurants, Mugaritz has gradually changed certain ingredients with a very important gastronomic value (fat, protein), for others equally valuable, but more healthy (herbs, cereals, tubers and vegetables).

This work, in terms of the interest of high level cooks in better understanding the impact of new procedures on a food's nutritional properties, is an important contribution to a new concept of gastronomy, focused not only on sensations but also on nutrition and health. These concepts could be equally applied in all restaurants and catering services to encourage changes in the distribution of dishes on the menus and understanding of the impact of some critical ingredients as added salt, sugar and fat (easily controllable) on the diet. The application of this knowledge at this level demands a better and wider transmission of dietetic and nutritional issues from school to professional education and training. Diet is the most important source of prevention of diseases such as obesity, diabetes and cardiovascular diseases, with high prevalence not only in the developed countries but around the world. Educational programs should take this into account as an important issue for maintenance of the next generation's nutritional and health status.

ACKNOWLEDGMENTS

This research was supported by the Centro para el Desarrollo Tecnológico e Industrial CDTI (Ministerio de industria, Turismo y Comercio, Spain) under the CENIT call (2009-2012). The project is entitled SENIFOOD (*Investigación industrial de dietas y alimentos con características específicas para las personas mayores*). The authors would also like to acknowledge the people from Mugaritz for their creativity and expertise in designing foods and for supplying the dishes.

REFERENCES

1. Association of Official Analytical Chemists International (AOAC), 2002. Association of Analytical Communities.
2. Bligh, E.G., Dyer, W.J., 1959. A rapid method of total lipid extraction and purification. Canadian Journal of Biochemistry and Physiology 37, 911–917.
3. European Commission, 1991a. Regulation (EC) 2568/91a. Characteristics of olive and olive pomace oils and their analytical methods. Official Journal of the European Communities and further revisions, Annex V, 81–90.
4. European Commission, 1991b. Regulation (EC) 2568/91b. Characteristics of olive and olive pomace oils and their analytical methods. Official Journal of the European Communities and Further Revisions, Annex X, B, L248, 44–45.

5. European Commission, 2006. Regulation (EC) No 1924/2006 of the European Parliament and of the Council of 20 December 2006 on nutrition and health claims made on foods. Official Journal of the European Union L12, 3–18.

6. European Commission, 2009. Working Document on the Setting of Nutrient Profiles. Preliminary Draft. Legal Proposal. European Commission, Health and Consumers Directorate-General.

7. European Commission, 2011. Regulation (EC) No. 1169/2011 of the European Parliament and of the Council of 25 October 2011 on the provision of food information to consumers amending Regulations (EC) No. 1924/2006 and (EC) No. 1925/2006 of the European Parliament and of the Council, and repealing Commission Directive 87/250/EEC, Council Directive 90/496/EEC, Commission Directive 1999/10/EC, Directive 2000/13/EC of the European Parliament and of the Council, Commission Directives 2002/67/EC and 2008/5/EC and Commission Regulation (EC) No. 608/2004.

8. European Commission, 2010. Regulation (EU) No 116/2010 of 9 February 2010 amending Regulation (EC) No 1924/2006 of the European Parliament and of the Council with regard to the list of nutrition claims. Official Journal of the European Union L37, 16–18.

9. European Food Safety Authority (EFSA), 2009a. Review of labelling reference intake values. The EFSA Journal 1008, 1–14.

10. European Food Safety Authority (EFSA), 2009b. Scientific opinion on the substantiation of health claims related to linoleic acid and maintenance of normal blood cholesterol concentrations (ID 489) pursuant to Article 13(1) of Regulation (EC) No 1924/2006. EFSA Journal 7, 1276.

11. European Food Safety Authority (EFSA), 2009c. Opinion on the substantiation of health claims related to alpha-linolenic acid and maintenance of normal blood cholesterol concentrations (ID 493) and maintenance of normal blood pressure (ID 625) pursuant to

Article 13(1) of Regulation (EC) No 1924/2006. EFSA Journal 7, 1252.

12. European Food Safety Authority (EFSA), 2010a. Scientific opinion on dietary reference values for fats, including saturated fatty acids, polyunsaturated fatty acids, monounsaturated fatty acids, trans fatty acids, and cholesterol. EFSA Journal 8 (3), 1461 2010, Carbohydrates and dietary fibre. EFSA Journal 8, 1462.

13. European Food Safety Authority (EFSA), 2010b. Scientific opinion on the substantiation of health claims related to protein and increase in satiety leading to a reduction in energy intake (ID 414, 616, 730), contribution to the maintenance or achievement of a normal body weight (ID 414, 616, 730), maintenance of normal bone (ID 416) and growth or maintenance of muscle mass (ID 415, 417, 593, 594, 595, 715) pursuant to Article 13(1) of Regulation (EC) No 1924/2006. EFSA Journal 8, 1811.

14. European Food Safety Authority (EFSA), 2010c. Scientific opinion on the substantiation of health claims related to calcium and maintenance of bones and teeth (ID 224, 230, 231, 354, 3099), muscle function and neurotransmission (ID 226, 227, 230, 235), blood coagulation (ID 230, 236), energy-yielding metabolism (ID 234), function of digestive enzymes (ID 355), and maintenance of normal blood pressure (ID 225, 385, 1419) pursuant to Article 13(1) of Regulation (EC) No 1924/ 2006. EFSA Journal 7, 1210.

15. European Food Safety Authority (EFSA), 2010d. Scientific opinion on the substantiation of health claims related to docosahexaenoic acid (DHA) and maintenance of normal (fasting) blood concentrations of triglycerides (ID 533, 691, 3150), protection of blood lipids from oxidative damage (ID 630), contribution to the maintenance or achievement of a normal body weight (ID 629), brain, eye and nerve development (ID 627, 689, 704, 742, 3148, 3151), maintenance of normal brain function (ID 565, 626, 631, 689, 690, 704, 742, 3148, 3151), maintenance of normal vision (ID 627, 632, 743, 3149) and maintenance of normal spermatozoa motility

(ID 628) pursuant to Article 13(1) of Regulation (EC) No 1924/2006. EFSA Journal 8, 1734.

16. European Food Safety Authority (EFSA), 2010e. Scientific opinion on the substantiation of health claims related to eicosapentaenoic acid (EPA), docosahexaenoic acid (DHA), docosapentaenoic acid (DPA) and maintenance of normal cardiac function (ID 504, 506, 516, 527, 538, 703, 1128, 1317, 1324, 1325), maintenance of normal blood glucose concentrations (ID 566), maintenance of normal blood pressure (ID 506, 516, 703, 1317, 1324), maintenance of normal blood HDL- cholesterol concentrations (ID 506), maintenance of normal (fasting) blood concentrations of triglycerides (ID 506, 527, 538, 1317, 1324, 1325), maintenance of normal blood LDL-cholesterol concentrations (ID 527, 538, 1317, 1325, 4689), protection of the skin from photo oxidative (UV-induced) damage (ID 530), improved absorption of EPA and DHA (ID 522, 523), contribution to the normal function of the immune system by decreasing the levels of eicosanoids, arachidonic acid-derived mediators and pro-inflammatory cytokines (ID 520, 2914), and "immunomodulating agent" (4690) pursuant to Article 13(1) of Regulation (EC) No 1924/2006. EFSA Journal 8, 1796.

17. European Food Safety Authority (EFSA), 2011. Scientific opinion on the substantiation of health claims related to foods with reduced amounts of sodium and maintenance of normal blood pressure (ID 336, 705, 1148, 1178, 1185, 1420) pursuant to Article 13(1) of Regulation (EC) No 1924/2006. EFSA Journal 9, 2237.

18. Lees, R., 1968. Laboratory Handbook of Methods of Food Analysis. London.

19. National Cholesterol Education Program (NECP), 2002. NECP expert panel on detection, evaluation, and treatment of high blood cholesterol in adults (adult treatment panel III. Final report). Circulation 106, 3143–3421

20. Vega, C., Ubbink, J., 2008. Molecular gastronomy: a food fad or science supporting innovative cuisine?. Trends in Food Science and Technology 19, 372–382.

INDEX